UNDERGRADUATE TEXTS IN CONTEMPORARY PHYSICS

Springer
New York
Berlin
Heidelberg
Hong Kong
London
Milan
Paris
Tokyo

UNDERGRADUATE TEXTS IN CONTEMPORARY PHYSICS

Cassidy, Holton, and Rutherford, Understanding Physics

Enns and McGuire, Computer Algebra Recipes: A Gourmet's Guide to the Mathematical Models of Science

Hassani, Mathematical Methods: For Students of Physics and Related Fields

Hassani, Mathematical Methods Using *Mathematica®:* For Students of Physics and Related Fields

Holbrow, Lloyd, and Amato, Modern Introductory Physics

Möller, Optics: Learning by Computing, with Examples Using Mathcad®

Roe, Probability and Statistics in Experimental Physics, Second Edition

Rossing and Chiaverina, Light Science: Physics and the Visual Arts

MATHEMATICAL METHODS USING *MATHEMATICA*®

For Students of Physics and Related Fields

Sadri Hassani

With 93 Illustrations and a CD-ROM

Springer

Sadri Hassani
Campus Box 4560
Department of Physics
Illinois State University
Normal, IL 61790-4560
USA
hassani@phy.ilstu.edu

COVER ILLUSTRATION: Gradient or differentiation with respect to distance is shown in two dimensions; the surface represents a function of x and y; the gradient is a vector *in the xy-plane*.

Library of Congress Cataloging-in-Publication Data
Hassani, Sadri.
Mathematical methods using *Mathematica*: for students of physics and related fields/ Sadri Hassani.
p. cm. — (Undergraduate texts in comtemporary physics)
Includes bibliographical references and index.
ISBN 0-387-95523-2 (softcover: alk. paper)
1. Physics—Mathematical models. 2. Mathematical physics—Data processing. 3. *Mathematica* (Computer file) I. Title. II. Series.
QC20 .H393 2002
530′.0285′53042—dc21 2002070732

ISBN 0-387-95523-2 Printed on acid-free paper.

Printed in the United States of America.

9 8 7 6 5 4 3 2 1 SPIN 10881953

Typesetting: Pages created by the author in LaTeX 2e using Springer's svsing2e.sty macro.

www.springer-ny.com

Springer-Verlag New York Berlin Heidelberg
A member of BertelsmannSpringer Science+Business Media GmbH

To my wife, Sarah,

and to my children,

Dane Arash and Daisy Bita

Preface

Over two years have passed since the publication of *Mathematical Methods*, my undergraduate textbook to which the present book was to be a companion. The initial motivation for writing this book was to take some examples from *Mathematical Methods* in which to illustrate the use of a symbolic language such as *Mathematica*®. However, after writing the first few pages, I realized very quickly that, for the book to be most effective, I had to go beyond the presentation of examples. I had to talk about the theory of numerical integration, discrete differentiation, solution of differential equations, and a number of other topics; thus the delay in the publication of the book.

As a result, the book has become a *self-contained* introduction to the use of computer algebra—specifically, *Mathematica*—for undergraduates in physics and related fields. Although many of the examples discussed here are taken from *Mathematical Methods*, no prior knowledge of the content of that book is essential for learning the *techniques* of computer algebra. Of course, a deeper understanding of the underlying physical ideas requires reading the relevant sections of *Mathematical Methods* or a book like it. For those interested in the underlying theories of the examples being discussed, I have placed the appropriate page (or section) numbers in the margin.

I have to emphasize that the book does not discuss programming in *Mathematica*. Nor does it teach all the principles and techniques of the most elegant utilization of *Mathematica*. The book can best be described as "learning the essentials of *Mathematica* through examples from undergraduate physics." In other words, *Mathematica* commands and techniques are introduced as the need arises.

I believe that some understanding of the theory behind the numerical calculations is important, especially if it can invoke some *Mathematica* usage. Therefore, I have included an entire chapter on the theory of the numerical solutions of differential equations, and a rather lengthy discussion on the theory behind numerical integration. In both discussions I make use of *Mathematica* to enhance the understanding of the theories.

After introducing the essential *Mathematica* commands in Chapter 1, I introduce vectors—using the calculation of electric fields and potentials of discrete charge distributions—and matrices—using the calculation of normal modes of mass-spring systems—in Chapter 2 as they are used in *Mathematica*. Chapter 3 discusses numerical integration and a variety of its applications in different physical settings such as the evaluation of electric, magnetic, and gravitational fields of various sources. Infinite series and finite sums are the subject of Chapter 4, in which the theory of numerical integration is used as a nice example of the use of summation in *Mathematica*. Chapter 5 is devoted entirely to a theoretical treatment of the numerical solution of differential equations, discussing such techniques as the Euler methods, the Runge–Kutta method, and the use of discrete differentiation in solving eigenvalue problems. In Chapter 6, I have chosen some examples from classical and quantum mechanics to illustrate how *Mathematica* solves ordinary differential equations.

This book can be used in conjunction with any undergraduate mathematical physics book. Many problems are inherently interesting but cannot be solved analytically. Once the student learns the theory and formal mathematics behind a concept and solves a number of simple and ideal examples analytically, he or she ought to be exposed to problems arising from real-world applications. *Mathematica* (or any other computer-algebra software) can be of tremendous help in treating such problems and exhibiting their solutions graphically (or otherwise). However, *Mathematica* has its greatest impact on the process of learning only if the student has completed the preliminary stage of deeply understanding the analytical methods of solution.

This is hardly the place to enter into the controversy surrounding the role of content and memorization in learning. However, as an educator witnessing the alarming rate at which calculators and computer-algebra software are substituting the learning of physics and mathematics, I feel obligated to emphasize the distinction between the *real utility* of technology and its *advertised glamour*. Technology can be a great tool of learning and teaching once students acquire a certain degree of mathematical maturity. And this maturity can be obtained only through a rigorous training in conventional mathematics that emphasizes content at all levels of a student's education.

The neglect of content—such as the multiplication table at the elementary level, and algebraic/trigonometric identities at the high school level—can have a detrimental effect on the mathematical and analytical ability of the pupil's mind. If the educators sequentially postpone the "memorization" of the multiplication table, algebraic and trigonometric identities, and differentiation and integration rules, arguing that such "facts" are always available on calculators and computers, then students will develop the skill of "pushing buttons" beautifully but will be incapable of doing the simplest integration. Some educators argue that lack of ability to multiply, integrate, or simplify an algebraic expression is not a drawback as long as there are calculators to do the job. To this I have to respond that heavy reliance on calculating machines does to the mind what heavy reliance on vehicular machines does to the body: it makes the mind lazy and inactive. Our minds need raw data—in the form of numbers and symbols in conjunction with the rules that manipulate them—to develop. A mind without data is like a symphony without notes, an opera without lyrics, a poem without words. I sincerely hope that the readers and users of this book will take this advice to heart.

Sadri Hassani
Campus Box 4560
Department of Physics
Illinois State University
Normal, IL 61790-4560, USA

e-mail: hassani@phy.ilstu.edu

Note to the Reader

I should point out from the very beginning that, as powerful as *Mathematica* is, it is only a tool. And a tool is more useful if its user has thought through the details of the task for which the tool is designed. Just as one needs to master multiplication—both conceptually (where and how it is used) and factually (the multiplication table)—before a calculator can be of any use, so does one need to master algebra, calculus, trigonometry, differential equations, etc., before *Mathematica* can be of any help. In short, *Mathematica* cannot think for you.

Mathematica, like any other calculational tool, is only as smart as its user can make it!

Once you have learned the concepts behind the equations and *know how to set up a specific problem*, *Mathematica* can be of great help in solving that problem for you. This book, of course, is not written to help you set up the problems; for that, you have to refer to your physics or engineering books. The purpose of this book is to familiarize you with the simple—but powerful—techniques of calculation used to solve problems that are otherwise insoluble. I have taken many examples from your undergraduate courses and have used a multitude of *Mathematica* techniques to solve those problems.

I encourage you to explore the CD-ROM that comes with the book. Not only does it contain all the codes used in the book, but it also gives many explanations and tips at each step of the solution of a problem. The CD-ROM is compatible with both *Mathematica* 3.0 and *Mathematica* 4.0.

Contents

1

Mathematica in a Nutshell

Mathematica® is a high-level computer language that can perform symbolic, numerical, and graphical manipulation of mathematical expressions. In this chapter we shall learn many of the essential *Mathematica* commands.

1.1 Running *Mathematica*

Installing and running *Mathematica* differ from one computer system to another. However, the heart of *Mathematica*, where the calculations are performed, is the same in all systems. *Mathematica* has two major components, the **kernel** and the **front end**. The front end is the window in which you type in your commands. These windows are generally part of **notebooks**, which are *Mathematica*'s interface with the kernel. The kernel is where the commands are processed. It could reside in the computer where the front end resides, or it could be in a remote computer.

of **kernels**, front ends, and notebooks

Mathematica is launched by double-clicking on its icon—or any other shortcut your computer system recognizes. Almost all front ends now incorporate notebooks, and I assume that the reader is communicating with *Mathematica* through this medium. The window of a notebook looks like any other window. After typing in your command, hold down the `Shift` key while hitting the `Return` key to execute that command. In Macintosh, the numeric `Enter` key will also do the job.

Shift+Return or numeric **Enter** tells *Mathematica* to start.

When you enter a command, *Mathematica* usually precedes it with an input sign such as `In[1]:=`; and when it gives out the result of the calculation, the output sign `Out[1]=` appears in front of the answer. The input and output numbers change as the session progresses. This is a convenient way of keeping track of all inputs and outputs for cross referencing. Thus, if you type `2+2` and enter the result, *Mathematica* turns it into `In[1]:= 2+2` and gives the result as `Out[1]= 4`.

In the remaining part of this chapter (and indeed throughout the book), we are going to discuss most of the commands an average user of *Mathematica* will need. Nevertheless, for the important details omitted in this book, the reader is urged to make frequent use of the definitive *Mathematica Book* [Wolf 96] as well as the **Help** menu, which includes the online version of the *Mathematica Book*.

1.2 Numerical Calculations

Mathematica recognizes two types of numerical calculations: integer and floating point. When the input of a mathematical expression is in integer form, *Mathematica*—unless asked specifically—does not approximate the final answer in decimal format. Consider asking *Mathematica* to add 11/3 to 217/43, by typing in

Mathematica calculates integer expressions exactly.

```
11/3 + 217/43
```

and pressing the numeric **Enter** key. *Mathematica* will give the answer as $\frac{1124}{129}$. On the other hand, if you type in

```
11./3 + 217/43
```

you will get the answer 8.71318. The difference between the two inputs is the occurrence of the floating (or decimal) point. In the second input *Mathematica* treats `11.` and all the other numbers in the expression as real numbers and manipulates them as such.

When *Mathematica* encounters expressions involving integers, it evaluates them and often gives the exact result. For example, for 5^{100}, *Mathematica* gives an exact 70-digit answer:

```
In[2]:= 5^100

Out[2]= 7888609052210118054117285652827862296732064351090\
        2300477027893066406250
```

use of `// N`

One can always get an approximate decimal answer by ending the input with `// N`

```
In[3]:= 5^100 // N

Out[3]= 7.88861 × 10^69
```

An alternative way of getting approximations is to use `N[expr,n]`, which returns the numerical value of `expr` to `n` significant figures. Here is how to find the numerical value of π to any desired significant figures. Simply replace 40 with some other (positive) integer:

use of **N[,]**

```
In[4]:= N[Pi,40]
```

Out[4]= 3.141592653589793238462643383279502884197

The example above illustrates how *Mathematica* denotes the constant π. Other mathematical constants also have their own notations:

`Pi`	$\pi = 3.14159$
`E`	$e = 2.71828$
`I`	$i = \sqrt{-1}$
`Infinity`	∞
`Degree`	$\pi/180$: as in 30 Degrees

Mathematica understands the usual mathematical functions with two caveats:

Functions begin with capital letters; arguments are enclosed in square brackets.

- All *Mathematica* functions begin with a capital letter.
- The arguments are enclosed in square brackets.

Here is a list of some common functions:

`Sqrt[x]`	$\sqrt{x}$	`Sin[x], ArcSin[x]`	sine, its inverse
`Exp[x]`	e^x	`Cos[x], ArcCos[x]`	cosine, its inverse
`Abs[x]`	absolute value	`Tan[x], ArcTan[x]`	tangent, its inverse
`n!`	the factorial	`Cot[x], ArcCot[x]`	cotangent, its inverse
`Log[x]`	natural log	`Log[b,x]`	log to base b

By default, the arguments of the trigonometric functions are treated as radians. You can, however, use degrees:

Arguments of the trigonometric functions are treated as radians.

```
In[5]:= Sin[Pi/3]-Cos[45 Degree]
```

Out[5]= $-\frac{1}{\sqrt{2}} + \frac{\sqrt{3}}{2}$

Note that *Mathematica* returns the *exact* result. This is because no floating point appeared in the arguments of the functions. Changing `3` to `3.` or `45` to `45.` returns `0.158919`. Similarly, `Sqrt[2]` will return $\sqrt{2}$, but

```
In[6]:= N[Sqrt[2],45]
```

Out[6]= 1.41421356237309504880168872420969807856967188

Reusing the existing expressions

Mathematica has a very useful shortcut for reusing the existing expressions.

`%`	the last result generated
`%%`	the next-to-last result generated
`%n`	the result on output line `Out[n]`

Typing in `%^2` squares the last result generated and returns its value. Similarly, `Sqrt[%6]` takes the square root of the result on output line `Out[6]`.

1.3 Algebraic and Trigonometric Calculations

The most powerful aspect of *Mathematica* is its ability to handle symbolic mathematics, including all the manipulations one encounters in algebra. The following is a partial list of algebraic expressions frequently encountered in calculations.

`Expand[expr]`	multiply products and powers in *expr*
`Factor[expr]`	write *expr* as products of minimal factors
`Simplify[expr]`	simplify *expr* (standard)
`FullSimplify[expr]`	simplify *expr* (comprehensive)
`PowerExpand[expr]`	transform $(xy)^p$ to $x^p y^p$; useful for changing $\sqrt{a^2}$ to a

Mathematica allows a convenient method of substituting values for a quantity in an expression:

`expr /. x -> value`	replace x by *value* in *expr*
`expr /. {x -> xval, y -> yval}`	perform several replacements

Here is an example of the use of some of the above:

```
In[1]:= x^2-2x+1 /. x -> 2 + y
```

Out[1]= $1 - 2(2 + y) + (2 + y)^2$

```
In[2]:= Expand[%]
```

Out[2]= $1 + 2y + y^2$

In[3]:= `Factor[%]`

Out[3]= $(1 + y)^2$

As "smart" as *Mathematica* is, it is too ignorant to do some of the most obvious things. You will have to ask it to do it. Consider the following:

In[4]:= `g=x^2 + y^2`

Out[41]= $x^2 + y^2$

In[5]:= `g/. {x->Cos[t],y->Sin[t]}`

Out[5]= $\text{Cos}[t]^2 + \text{Sin}[t]^2$

In[6]:= `TrigReduce[g]`

Out[6]= $x^2 + y^2$

In[7]:= `TrigReduce[%5]`

Out[7]= 1

This example illustrates a number of *Mathematica* subtleties that are worth mentioning at this point. First, note that the substitution `x->Cos[t]` and `y->Sin[t]` did not change the value of `g`, as evident in `Out[6]`. Second, *Mathematica* does not automatically "remember" even the simplest trigonometric identity, if you do not "remind" it of the identity. Third, *Mathematica* has some trigonometric "reminders," some of which are gathered below:

some trigonometric commands

`TrigExpand[expr]`	expand the trig *expr* into a sum of terms
`TrigFactor[expr]`	write the trig *expr* as products of terms
`TrigReduce[expr]`	simplify trig *expr* using trig identities

Here is an illustration of how these work:

In[8]:= `TrigExpand[Sin[x+y]]`

Out[81]= $\text{Cos}[y]\,\text{Sin}[x] + \text{Cos}[x]\,\text{Sin}[y]$

In[9]:= `h=% /. {y->2x}`

Out[9]= $\text{Cos}[2x]\,\text{Sin}[x] + \text{Cos}[x]\,\text{Sin}[2x]$

```
In[10]:= TrigExpand[h]
```

Out[10]= $3\,\mathrm{Cos}[x]^2\,\mathrm{Sin}[x] - \mathrm{Sin}[x]^3$

```
In[11]:= TrigFactor[%]
```

Out[11]= $(1 + 2\,\mathrm{Cos}[2x])\,\mathrm{Sin}[x]$

```
In[12]:= TrigReduce[%]
```

Out[12]= $\mathrm{Sin}[3x]$

For most purposes, the commands discussed so far are adequate. However, sometimes—especially with rational expressions—other commands may come in handy. The following are examples of such commands:

`Factor[expr]`	reduce *expr* to a product of factors
`Together[expr]`	put all terms of *expr* over a common denominator
`Apart[expr]`	separate *expr* into terms with simple denominators
`Cancel[expr]`	cancel common factors in numerator and denominator of *expr*

Let us look at an example. Consider the rational expression

```
In[1]:= f=(2x^5-x^3+6x^2-x+3)/(6x^3-4x^2+3x-2)
```

Out[1]:=

$$\frac{3 - x + 6x^2 - x^3 + 2x^5}{-2 + 3x - 4x^2 + 6x^3}$$

First we separate f into terms with simple denominators:

```
In[2]:= g=Apart[f]
```

Out[2]:=

$$-\frac{5}{27} + \frac{2x}{9} + \frac{x^2}{3} + \frac{71}{27(-2 + 3x)}$$

Then we put them back together again:

```
In[3]:= Together[g]
```

Out[3]:=

$$\frac{3 - x + x^3}{-2 + 3x}$$

This is not f because, in the process of putting g together, *Mathematica* simplified the expression, canceling out the common factors in the numerator and denominator. To see this, we reproduce f by typing it in:

```
In[4]:= f
```

```
Out[4]:=
```

$$\frac{3 - x + 6x^2 - x^3 + 2x^5}{-2 + 3x - 4x^2 + 6x^3}$$

and ask *Mathematica* to cancel common factors in its numerator and denominator:

```
In[5]:= Cancel[f]
```

```
Out[5]:=
```

$$\frac{3 - x + x^3}{-2 + 3x}$$

The following are some useful commands with which one can separate different parts of an expression:

`Numerator[expr]`	numerator of *expr*
`Denominator[expr]`	denominator of *expr*
`Part[expr,n]`	*n*th term of *expr*
`Coefficient[expr,form]`	coefficient of *form* in *expr*

As an example, consider

```
In[1]:= u=((x^2-2y+3)^4 Tan[x])/(x+Sin[3x] Cos[2y])
```

```
Out[1]:=
```

$$\frac{(x^2 - 2y + 3)^4 \text{ Tan[x]}}{x + \text{ Cos[2y] Sin[3x]}}$$

Now isolate the numerator:

```
In[2]:= num=Numerator[u]
```

```
Out[2]:=
```

$$(x^2 - 2y + 3)^4 \text{ Tan[x]}$$

Tell *Mathematica* to produce the (obvious) coefficient of `Tan[x]`:

```
In[3]:= Coefficient[num, Tan[x]]
```

```
Out[3]:=
```

$$(x^2 - 2y + 3)^4$$

That was easy. But the following is not!

```
In[4]:= Coefficient[%, y^2]

Out[4]:=
```

$$216 + 144x^2 + 24x^4$$

Now type in

```
In[5]:= TrigExpand[Denominator[u]]

Out[5]:=
```

$$x + 3\ \mathrm{Cos[x]}^2\ \mathrm{Cos[y]}^2\ \mathrm{Sin[x]} - \mathrm{Cos[y]}^2\ \mathrm{Sin[x]}^3$$
$$- 3\ \mathrm{Cos[x]}^2\ \mathrm{Sin[x]}\ \mathrm{Sin[y]}^2 + \mathrm{Sin[x]}^3\ \mathrm{Sin[y]}^2$$

To find the coefficient of $\sin x$ type in

```
In[6]:= Coefficient[%, Sin[x]]

Out[6]:=
```

$$3\ \mathrm{Cos[x]}^2\ \mathrm{Cos[y]}^2 - 3\ \mathrm{Cos[x]}^2\ \mathrm{Sin[y]}^2$$

1.4 Calculus in *Mathematica*

The ability to combine algebraic, trigonometric, and analytic calculations makes *Mathematica* an extremely powerful tool. It has the following commands for differentiation of a function:

D[f, x]	the (partial) derivative $\dfrac{\partial f}{\partial x}$
D[$f, x_1, x_2, \ldots$]	the multiple partial derivative $\dfrac{\partial}{\partial x_1}\dfrac{\partial}{\partial x_2}\cdots f$
D[$f, \{x, n\}$]	the nth derivative $\dfrac{\partial^n f}{\partial x^n}$

When the function depends on a single variable, the following abbreviations—common in mathematical literature—can be used:

f'[x]	the derivative $f'(x)$
f''[x]	the second derivative $f''(x)$
f''''''''[x]	the eighth derivative $f^{(8)}(x)$

Mathematica can differentiate simple expressions

```
In[1]:= D[x^2,x]
```

Out[1]= $2x$

as easily as some not-so-simple ones:

```
In[2]:= D[Sin[x*y^2/Exp[x+y]],x,y]
```

Out[2]= $(2E^{-x-y}y-2E^{-x-y}xy-E^{-x-y}y^2+E^{-x-y}xy^2)Cos[E^{-x-y}xy^2]$
$-(2E^{-x-y}xy-E^{-x-y}xy^2)(E^{-x-y}y^2-E^{-x-y}xy^2)Sin[E^{-x-y}xy^2]$

Mathematica can perform definite and indefinite integration using its enormous table of integrals as well as powerful internal routines. The symbol for integration commands are very intuitive:

`Integrate[`f, x`]`	the indefinite integral $\int f\,dx$
`Integrate[`$f, \{x, a, b\}$`]`	the definite integral $\int_a^b f\,dx$

Mathematica knows the elementary indefinite integrals

```
In[3]:= Integrate[x^2 Log[x],x]
```

Out[3]= $-\frac{x^3}{9}+\frac{1}{3}x^3 Log[x]$,

and some famous definite integrals:

```
In[4]:= Integrate[Exp[-x^2],{x,0,Infinity}]
```

Out[4]= $\frac{\sqrt{\pi}}{2}$.

It is familiar enough with some famous functions to give the result of some integrations in terms of them:

```
In[5]:= Integrate[Sqrt[1-k Sin[x]^2],x]
```

Out[5]= `EllipticE[x,k]`.

However, it cannot evaluate all integrals analytically:

```
In[6]:= Integrate[x^x],{x,0,1}]
```

Out[6]= $\int_0^1 x^x\,dx$

Nevertheless, it can always give you the numerical value of the integral to any desired accuracy:

```
In[7]:= N[%,20]

Out[7]= 0.78343051071213440071
```

defining functions (or delayed assignments)

A powerful tool of *Mathematica*, which the reader will find extremely useful, is its ability to manipulate user-defined functions. These functions can be not only in terms of the internal built-in functions, but also in terms of procedures. We define a function f by typing `f[x_]:=` and putting the desired expression, formula, function, etc. involving the "dummy" variable x on the right-hand side of the equation. For example, we define f as follows:[1]

```
In[1]:= f[x_]:= a x^3 Sin[a x]
```

This teaches *Mathematica* a rule by which it takes the argument of f and manipulates it according to the instructions given on the right-hand side. Notice the important _ (referred to as "blank") on the left-hand side. The colon to the left of the equal sign is not significant, as it only suppresses the output of that line. We could replace the argument of the function so defined with any expression, and the function gets evaluated accordingly.

```
In[2]:= f[b]
```

Out[2]= $ab^3\,\mathrm{Sin}[ab]$

```
In[3]:= f[a]
```

Out[3]= $a^4\,\mathrm{Sin}[a^2]$

```
In[4]:= f[Sin[y]]
```

Out[4]= $a\,\mathrm{Sin}[y]^3\,\mathrm{Sin}[a\,\mathrm{Sin}[y]]$

Once the function is defined, many operations such as differentiation and integration can be performed on it. For example, we can differentiate the complicated function $f(\sin y)$ shown in `Out[4]` above:

```
In[5]:= D[f[Sin[y]],y]
```

Out[5]= $a^2\,\mathrm{Cos}[y]\,\mathrm{Cos}[a\,\mathrm{Sin}[y]]\,\mathrm{Sin}[y]^3+3a\,\mathrm{Cos}[y]\,\mathrm{Sin}[y]^2\,\mathrm{Sin}[a\,\mathrm{Sin}[y]]$

Or we can integrate the same function:

---

[1]Note that a space between two symbols is the same as multiplication of those symbols. Thus, `a x` is equivalent to `a*x`.

```
In[6]:= Integrate[f[Sin[y]],{y,0,1}]
```

Out[6]= $a\int_0^1 \text{Sin}[y]^3\,\text{Sin}[a\,\text{Sin}[y]]\,dy$

So, *Mathematica* did not know how to integrate the function. But, as always, it can easily compute it numerically. To do so, we need first to assign a numerical value to a:

```
In[7]:= % /. a->2
```

Out[7]= $2\int_0^1 \text{Sin}[y]^3\,\text{Sin}[2\,\text{Sin}[y]]\,dy$

```
In[8]:= N[%]
```

Out[8]= 0.340467

As mentioned earlier, functions can be procedures. A frequently occurring procedure is summation of related terms. *Mathematica* has a built-in expression for summation:

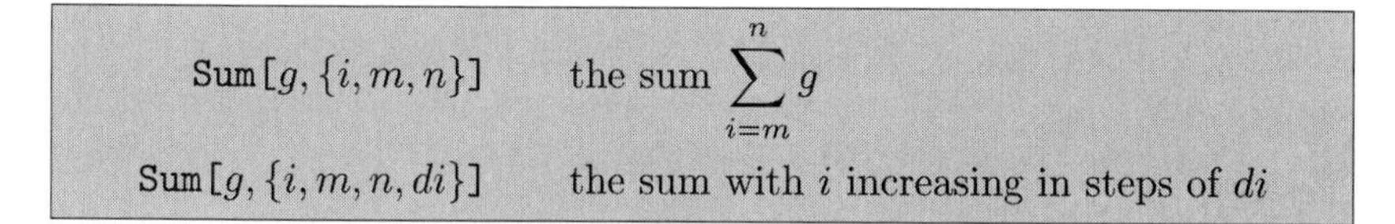

`Sum`$[g, \{i, m, n\}]$	the sum $\sum_{i=m}^{n} g$
`Sum`$[g, \{i, m, n, di\}]$	the sum with i increasing in steps of di

Consider the familiar sum

```
In[1]:= Sum[1/k!,{k,0,5}]
```

Out[1]= $\frac{163}{60}$

```
In[2]:= N[%]
```

Out[2]= 2.71667

Notice two things: one, that *Mathematica* is familiar with the convention $0! = 1$; and two, that it gives the result of the summation as a fraction. The sum above contains six terms. If we want to include more terms in the sum, we have to change 5 to some other number. For example,

```
In[3]:= Sum[1/k!,{k,0,20}]
```

Out[3]= $\frac{6613313319248080001}{2432902008176640000}$

```
In[4]:= N[%]
```

Out[4]= 2.71828.

A more economical way of handling this is to define a function whose argument is the number of terms. In fact, we can do better. We define a function with *two* arguments as follows:

```
In[5]:= Clear[f,x,n]

In[6]:= f[x_,n_]:= Sum[x/k!,{k,0,n}]
```

Use **Clear[f,x, ...]** to override previous assignment to $f, x, \ldots$.

Notice that k, being a dummy index, does not appear on the left-hand side. In fact, we can use any other symbol in place of k on the right-hand side with no noticeable consequence. The statement `Clear[f,x,n]` is a precaution to clear all the previous definitions of f, x, and n. The use of this statement is a good habit to get into to make sure that no values assigned previously to a variable or a function enter in the current definition of the function.

With `f[x,n]` at our disposal, we can now find the values of the sum, not only for $x = 1$, but for any x—and any n, of course. For instance,

`In[7]:= f[y,8]`

`Out[7]:=` $1 + y + \frac{y^2}{2} + \frac{y^3}{6} + \frac{y^4}{24} + \frac{y^5}{2} + \frac{y^2}{120} + \frac{y^7}{5040} + \frac{y^8}{40320}$;

or

`In[8]:= f[1,30]`

`Out[8]:=` $\frac{55464002213405654539818183437977}{20404066139399312202792960000000}$

`In[9]:= N[%,35]`

`Out[9]:=` 2.7182818284590452353602874713526624

`In[10]:= N[E,35]`

`Out[10]:=` 2.7182818284590452353602874713526625

`In[11]:= f[1,Infinity]`

`Out[11]:=` E.

`In[10]` asks *Mathematica* to give the numerical value of the base of natural logarithm to 35 significant figures. `Out[11]` shows that *Mathematica* is familiar with the infinite series expansion of the base of natural logarithm. In fact, it is more intelligent!

`In[12]:= f[2,Infinity]`

`Out[12]:=` E^2

```
In[13]:= f[y,Infinity]
```

Out[13]:= E^y.

At the beginning of Section 1.3, we mentioned *Mathematica*'s substitution rule `x-> value`. This same rule applies to functions in a more general and powerful way. Here are some examples:

```
In[1]:= f[t]+f[s]/. t->-5, f[s]->Sin[y]
```

Out[1]:= $f[-5] + \mathrm{Sin}[y]$

```
In[2]:= f[t]+f[s]/. f[x_]->x^2
```

Out[2]:= $s^2 + t^2$

```
In[3]:= f[t]+f[s]/. f[x_]->f[x y]
```

Out[3]:= $f[ty] + f[sy]$

```
In[4]:= % /. f[a_ b_]->f[a]+f[b]
```

Out[4]:= $f[t] + f[s] + 2f[y]$.

1.5 Numerical Mathematics

Mathematica is not only a powerful program for symbolic mathematics, it is also capable of handling sophisticated numerical calculations. In fact, almost all the symbolic operations have a numeric counterpart, distinguished from the former by the concatenation of the letter `N` to the name of the symbolic operation.

One of the most common problems encountered in numerical mathematics is solving equations. Equations in *Mathematica* are described by a double equal sign. A single equal sign is an assignment, not an equation. The relevant statements are

For equations use == not =.

`Solve[lhs==rhs,x]`	solve the single equation for x
`NSolve[lhs==rhs,x]`	numerically solve the single equation for x
`expr /. sol`	evaluate *expr* using the values obtained in *sol*

`Solve` tries to solve the equation $lhs == rhs$, where lhs and rhs are expressions in x, and give you the exact solution in the form of a list; it can handle mostly polynomials of degree 4 or less. `NSolve` does the same numerically; it can handle polynomials of all degrees. You can assign a label to the `Solve` or `NSolve` expressions—we have designated this label as *sol* above—and use that label to evaluate expressions involving x at the roots of the equations. Here is a familiar example:

```
In[1]:= Solve[a x^2 +b x +c == 0,x]
```

Out[1]:=

$$\left\{\left\{x \to \frac{-b-\sqrt{b^2-4ac}}{2a}\right\},\left\{x \to \frac{-b+\sqrt{b^2-4ac}}{2a}\right\}\right\}$$

Solution of equations is put out in the form of a list.

As seen above, *Mathematica* gives solutions in the form of a list designated as entries in a curly bracket separated by commas. The solutions are not given as simple entries, but as a "replacement rule"—again in curly brackets. This is especially convenient when the solutions are to be substituted in other expressions.

Mathematica can also solve cubic and quartic equations. A well-known result in mathematics states, however, that higher-degree polynomials cannot be solved analytically in terms of radicals. Nevertheless, *Mathematica* can find the roots numerically.

As another example, suppose we want to find the value of `f[x]` at the roots of the polynomial $1 + x + 2x^2 + 3x^3 + 4x^4 + 5x^5$. First we find the (complex) numerical roots of the polynomial and call them *sol*:

```
In[1]:= sol=NSolve[1 + x + 2 x^2 +3 x^3 +4 x^4
          + 5 x^5 == 0,x]
```

Out[1]:=

$$\{\{x \to -0.789728\},\{x \to -0.326925 - 0.674868I\}$$
$$\{x \to -0.326925 + 0.674868I\}\{x \to 0.321789 - 0.588911I\}$$
$$\{x \to 0.321789 + 0.588911I\}\}$$

Then we evaluate the function at the roots:

```
In[2]:= f[x]/. sol
```

Out[2]:=

$$\{f[-0.789728], f[-0.326925 - 0.674868I]$$
$$f[-0.326925 + 0.674868I] f[0.321789 - 0.588911I]$$
$$f[0.321789 + 0.588911I]\}$$

Mathematica can solve a simultaneous set of equations involving several variables. The rules are the same as before, except that—as is common in *Mathematica*—one lists the equations in one set of curly brackets separated by commas and the unknowns in another set of curly brackets. For example,

```
In[3]:= h=NSolve[{x^5 -y^5==1,x -2 y == 1} ,{x,y}]
```

yields

Out[3]:=

$$\{\{x \to -0.902376 - 0.430949I, y \to -0.951188 - 0.215474I\},$$
$$\{x \to -0.902376 + 0.430949I, y \to -0.951188 + 0.215474I\},$$
$$\{x \to 0.321731 - 0.946831I, y \to -0.339134 - 0.473416I\},$$
$$\{x \to 0.321731 + 0.946831I, y \to -0.339134 + 0.473416I\},$$
$$\{x \to 1., y \to 0.\}\}$$

The solutions, as expected, are complex numbers. We can easily find the absolute values of these complex numbers. Note that the set of solutions is labeled h in this case:

```
In[4]:= {Abs[x], Abs[y]} /. h
```

Out[4]:=

$$\{\{1., 0.975289\}, \{1., 0.975289\}, \{1., 0.582352\},$$
$$\{1., 0.582352\}, \{1., 0.\}\}$$

It is interesting that the absolute values of x are all equal to 1.

Mathematica cannot `Solve` a simple equation such as $\cos x = x$. It cannot even `NSolve` it. This is because the equation is not algebraic; it is a *transcendental* equation. There is, however, another command that can be used in such situations:

`FindRoot[lhs==rhs,{`x, x_0`}]`	find a numerical solution starting with x_0

Unfortunately, finding a root requires a prior estimate of the location of the root. The value x_0 provided to *Mathematica* should not be too far from the root of the equation, although in some cases *Mathematica* can solve the equation with x_0 far away from the root. The reason for this drawback is that *Mathematica* mostly uses Newton's method for finding roots. Let us briefly describe this method.

Newton's method of finding roots of equations

Suppose you want to find the roots of the equation $f(x) = 0$, i.e., the points at which f intersects the x-axis. Estimate one of the roots, say r, and pick a value x_0 close (it does not have to be too close!) to that root. Now approximate the function with the tangent line L_1 that passes through $(x_0, f(x_0))$. Find the point that the tangent line intersects the x-axis; call this x_1. Since the tangent line is given as a linear function involving x_0, we can always solve for x_1 entirely in terms of x_0. In fact, setting $y = 0$ in the equation of the tangent line

$$y - f(x_0) = f'(x_0)(x - x_0)$$

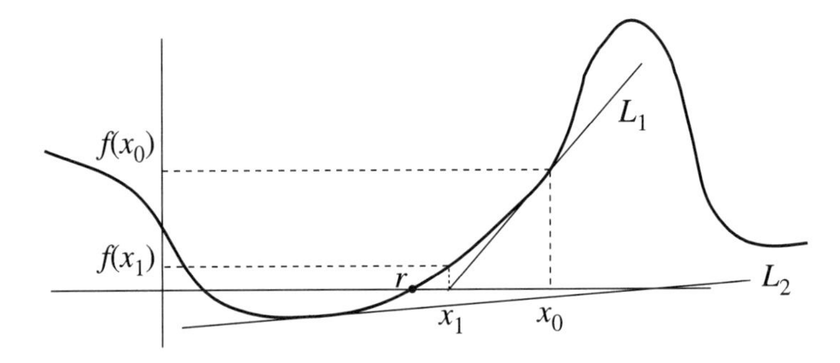

FIGURE 1.1. By approximating the function with tangent lines, we can get successively closer to the root of the function. But we are not always lucky! The tangent line of the starting point on the left of the root intersects the x-axis too far away from r.

finds the x-intercept, i.e., x_1. Thus,

$$0 - f(x_0) = f'(x_0)(x_1 - x_0) \Rightarrow x_1 = x_0 - \frac{f(x_0)}{f'(x_0)}$$

Generally, x_1 is closer to the root r than x_0, as shown in Figure 1.1, and can be used as the input into the above equation to give a closer value to r:

$$x_2 = x_1 - \frac{f(x_1)}{f'(x_1)}$$

and in general,

$$x_{n+1} = x_n - \frac{f(x_n)}{f'(x_n)} \tag{1.1}$$

In particular, if we are interested in the mth root of a, we use the function $f(x) = x^m - a$. This gives

$$x_{n+1} = x_n - \frac{x_n^m - a}{m x_n^{m-1}} = \left(1 - \frac{1}{m}\right) x_n + \frac{a}{m x_n^{m-1}} \tag{1.2}$$

Mathematica has a built-in function that is especially suited for this purpose:

`NestList[f,x,n]`	generates the list $\{x, f[x], f[f[x]], \ldots\}$ where f is nested up to n times deep

So, if we apply Equation (1.2) a large number of times, we get the mth root of a to any desired accuracy. Actually, the process converges very rapidly. For example, to find $\sqrt[3]{8}$, set $a = 8$, $m = 3$, and define

```
In[5]:= f[x_] := 2 x/3+8/(3 x^2)
```

and type in

```
In[6]:= NestList[f,1,6]
```

to obtain

```
Out[6]:= {1.,3.33333,2.46222,2.08134,2.00314,2.,2.}
```

We see that the last two iterations give the same result—up to *Mathematica*'s default precision. In reality, the fifth and the sixth iterations are 2.0000049 and 2.000000000012, respectively!

In the example above, the starting point, $x_0 = 1$, was actually close enough—and strategically located—to the root for the procedure to give a very good approximation in just six iterations. However, many times x_0 may be too far from the root r, or it may be so located relative to r that the point of intersection with the x-axis is farther than the original point. An example of such a situation is the point in Figure 1.1 to the left of r—close to an extremum of f—the intersection of whose tangent line L_2 with the x-axis may be so far away that no hope of ever getting closer to the intended root exists. Trigonometric and related functions, due to their multiplicity of extrema, exhibit this unwanted property, and Newton's method will not always work for them.

In such undesirable circumstances, Newton's method may take a large number of successive calculations to get to the root, or it may never reach the root. The default number of successive approximations for *Mathematica* is 15. If after 15 trials *Mathematica* does not get close enough to the root, it gives a warning and prints out the latest approximation. So, in using `FindRoot`, it is a good idea to have a rough estimate of the root and be ready to try a different estimate if the first one does not yield results.

In case of multiple roots, we can specify the interval in which the root may lie. This is done by using the following command:

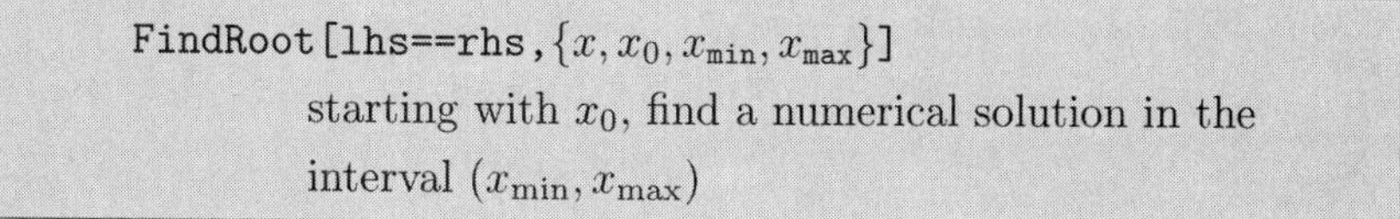
`FindRoot[lhs==rhs,`$\{x, x_0, x_{\min}, x_{\max}\}$`]` starting with x_0, find a numerical solution in the interval $(x_{\min}, x_{\max})$

The drawback of this command is that as soon as *Mathematica* reaches a value outside the interval it stops, even though it could have reached the desired accuracy had it continued. Here is an example:

```
In[1]:= FindRoot[Tan[x] == x,{x,7}]
```

```
Out[1]:= {x -> 7.72525}
```

But

```
In[2]:= FindRoot[Tan[x] == x,{x,7.5,7,8}]
```

produces the warning

```
FindRoot :: regex : Reached the point {8.15469} which
is outside the region {{7., 8.}}
```

and the value $\{x \to 8.15469\}$ is printed on the screen. It is, therefore, best to simply specify a good starting point for each root.

1.6 Graphics

The results of numerical mathematics—and many symbolic manipulations as well—are best conveyed in the form of graphics. *Mathematica* has a very powerful graphics capability suitable for simple two-dimensional plots as well as complex, multicolored three-dimensional images.

The following is a list of different graphics in *Mathematica*:

- Two-dimensional plots consisting of
  - ordinary plots of the form $y = f(x)$
  - parametric plots of the form $x = f(t), y = g(t)$
  - two-dimensional contour plots of surfaces
  - two-dimensional density plots of surfaces
- Three-dimensional plots consisting of
  - surface plots of the form $z = f(x, y)$
  - parametric plot of a curve: $x = f(t), y = g(t), z = h(t)$
  - parametric plot of a surface: $x = f(t, u), y = g(t, u), z = h(t, u)$

Let us briefly discuss each of these graphics separately.

1.6.1 *Simple Plots*

The simplest kind of plot is the two-dimensional graph of a function of the form $y = f(x)$. The command for this kind of plot is

`Plot[f[x],{x,a,b}]`	plot $f(x)$ from $x = a$ to $x = b$

Mathematica knows how to avoid infinities in graphs.

where `f[x]` can be either an internal function or a user-defined function involving internal functions and procedures. For example,

```
In[1]:= Plot[Cot[x],{x,-3,3}]
```

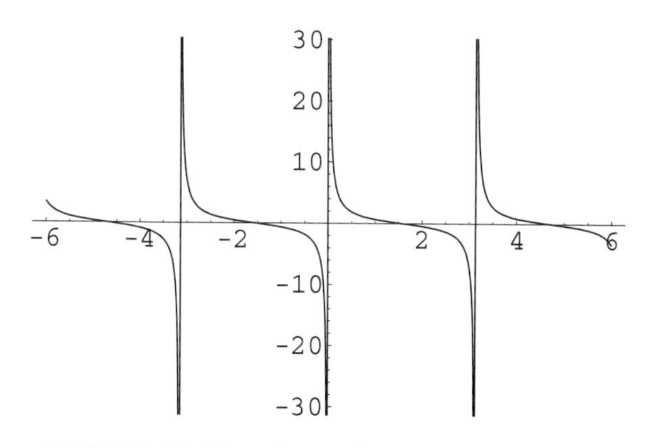

FIGURE 1.2. Plot of $\cot x$ from $x = -3$ to $x = 3$.

yields the plot of Figure 1.2. Note that *Mathematica* handles the singularities of the function nicely.

Now let us consider an example in which the function is user-defined. Take the function to be a simple integral:

```
In[2]:= f[x_]:= Integrate[E^(-t^2), {t,0,x}]
```

This—within a constant multiple—is the famous error function used extensively in statistics and probability, known to *Mathematica* as `Erf[x]`.[2] We then tell *Mathematica* to plot this function for us in some convenient interval:

```
In[3]:= Plot[f[x],{x,0,3}]
```

The output of this plot is shown in Figure 1.3.

We can plot several functions on the same graph. We simply put all functions in curly brackets separated by commas. For example,

plotting several functions on the same graph

```
In[4]:= Plot[{10 E^(-x^2), 20 Sin[x], Tan[x]},{x,-3,3}]
```

produces the plot of Figure 1.4. We can change the look of each plot—e.g., using dashes or different shades of gray—to differentiate between them more easily. For hints on changing the style of a plot, the reader may consult the *Mathematica Book* or the `Help` menu of *Mathematica*.

1.6.2 Parametric Plots

Functions of the form $y = f(x)$ can have only one value for y for a given x. So, a simple graph such as a circle of radius a can only be drawn by

[2]The fact that *Mathematica* recognizes the error function is immaterial here. We could use any (complicated) function as the integrand so that the result of integration is not known to *Mathematica*.

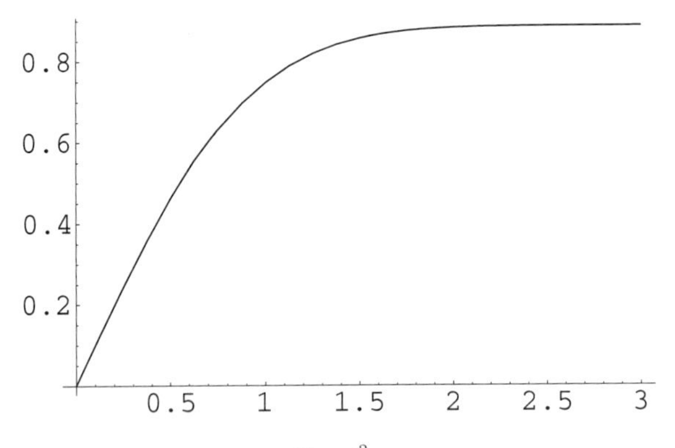

FIGURE 1.3. Plot of $\int_0^x e^{-t^2}\,dt$ from $x = 0$ to $x = 3$.

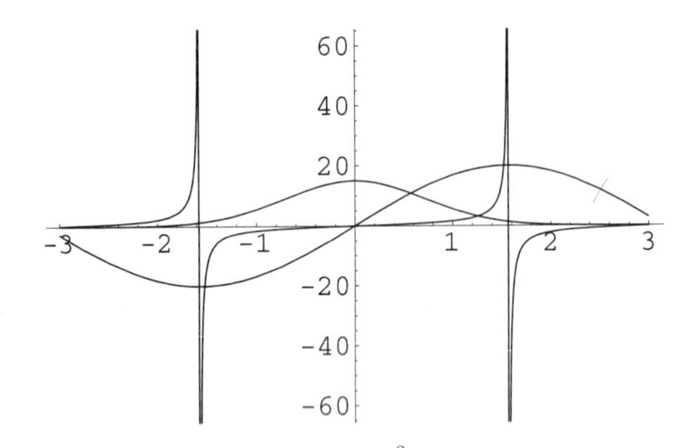

FIGURE 1.4. Plot of three functions, $15e^{-x^2}$, $20\sin x$, and $\tan x$ from $x = -3$ to $x = 3$.

juxtaposing the two half-circles $y = \sqrt{a^2 - x^2}$ and $y = -\sqrt{a^2 - x^2}$. There is, however, a more elegant way around this: using parametric equations of a curve. In fact, parametric equations can produce curves that are impossible with any number of curves each described by some $y = f(x)$.

The command for the parametric plot of $x = f(t), y = g(t)$ is, not surprisingly,

`ParametricPlot[{f[t],g[t]},{t,a,b}]`	make a parametric plot from $t = a$ to $t = b$

To produce our full circle mentioned above, we note that in *polar* coordinates its equation is $r = a$ and θ goes from zero to 2π. Therefore, $x = a\cos\theta$

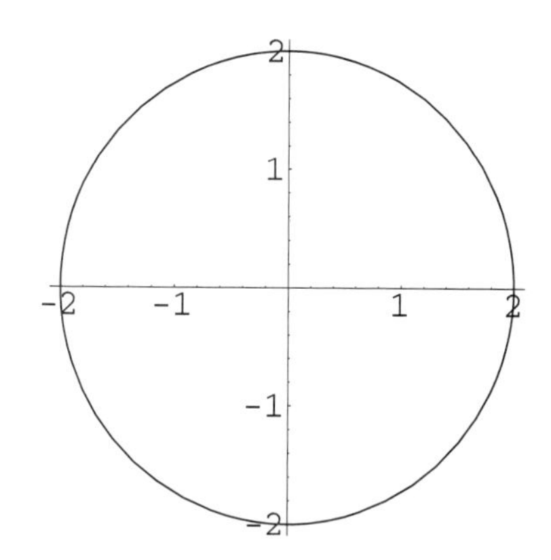

FIGURE 1.5. Parametric plot of a circle.

and $y = a\sin\theta$ can be interpreted as the parametric equations of the circle. Thus,

use of the option **AspectRatio**

```
In[5]:= ParametricPlot[{2 Cos[t], 2 Sin[t]},{t, 0, 2 Pi},
  AspectRatio->Automatic]
```

will produce a circle of radius 2. This circle is shown in Figure 1.5. The option `AspectRatio->Automatic` tells *Mathematica* to use the same units for the horizontal and vertical axes.

With a parametric plot you can produce all the pretty polar plots you learned in calculus.[3] For example, recall that $r = \sin(n\theta)$ produces a $2n$-leaved clover if n is even and an n-leaved clover if n is odd. Because $x = r\cos\theta$ and $y = r\sin\theta$, we can easily produce these clovers. Thus, to produce a 4-leaved clover, use

```
In[5]:= ParametricPlot[{Cos[t] Sin[2 t],
          Sin[t] Sin[2 t]},{t, 0, 2 Pi}]
```

On the other hand,

```
In[5]:= ParametricPlot[{Cos[t] Sin[5 t],
          Sin[t] Sin[5 t]},{t, 0, 2 Pi}]
```

will produce a 5-leaved clover. Both of these are shown in Figure 1.6. In producing this figure, we have used the following two *Mathematica* commands:

`GraphicsArray[{g1,g2, ... }]`	arranges several graphs in one row
`Show[g]`	redraws graph g

[3]Actually, *Mathematica* has a command called `PolarPlot`, which can directly plot graphs whose equations are given in polar form.

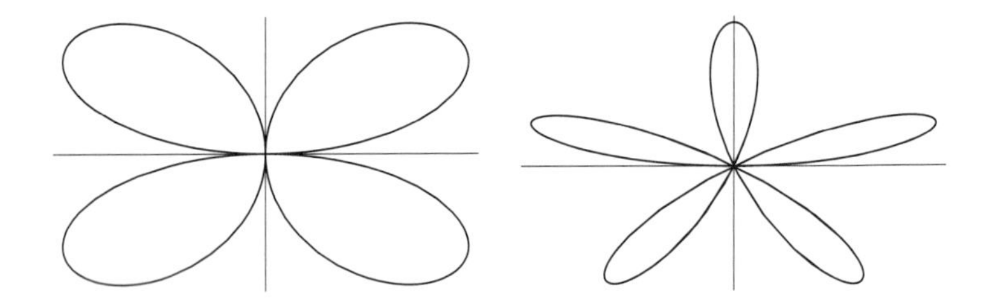

FIGURE 1.6. Drawing clovers using `ParametricPlot`.

and one graph *option*:

`Ticks -> None`	hide the tick marks

Options can customize your graph.

`Options` are properties that can be used in plots to customize their presentation. They are usually placed at the end of the `Plot` argument, separated by commas from each other and the rest of the argument.

1.6.3 Contour and Density Plots

The two-dimensional contour and density plots are less common in the introductory courses. However, they are indispensable in computationally intensive advanced courses and research areas. A contour plot of $f(x, y)$ is a topographical map of the function. It gives the boundaries of slices cut through the function by planes parallel to the xy-plane. The density plot depicts the *value* of f at a regular array of points. Higher regions are shown in lighter shades. The syntax for these plots, for which x runs from a to b, and y, from c to d, are

```
ContourPlot[f[x,y],{x,a,b},{y,c,d}]
DensityPlot[f[x,y],{x,a,b},{y,c,d}]
```

the option **ContourShading**

Figure 1.7 shows the contour plot (left) and the density plot (right) of the function e^{xy} for $-2 \leq x \leq 2$ and $-2 \leq y \leq 2$. Sometimes contour plots are drawn without shading. This can be done using the option `ContourShading->False`. However, unless the printer is incapable of handling gray levels well, it is actually preferable to leave the shading in, because the gray level gives information about the "height" of the function: the lighter the shade, the larger the value of the function.

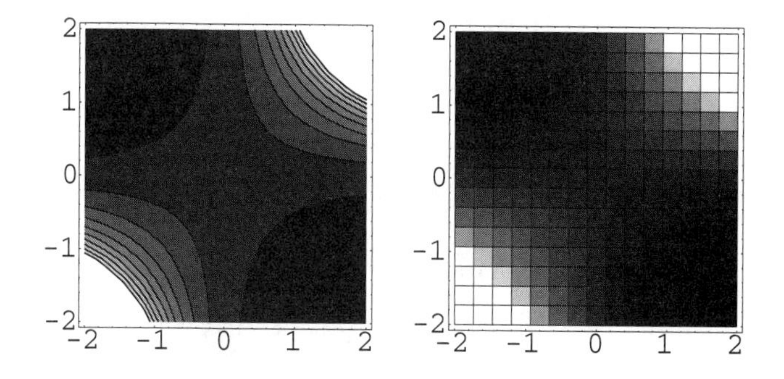

FIGURE 1.7. The contour plot (left) and the density plot (right) of e^{xy}.

1.6.4 Three-Dimensional Plots

Instead of showing only the two-dimensional contours or density variation of a three-dimensional function, one can show a three-dimensional plot of that function. The command is simple enough: if you want a plot of a function f with x between a and b and y between c and d, then type in

```
Plot3D[f[x,y],{x,a,b},{y,c,d}]
```

For example,

```
In[1]:= Plot3D[ Cot[t u],{t, -3, 3},{u, -3, 3}]
```

produces the graph of Figure 1.8. This is not entirely true! *Mathematica* produces a much rougher version of this graph. To obtain the graph of Figure 1.8 one has to change the default options so that the number of points on each axis as well as the range of the function (the range of the vertical axis) is larger than the defaults. (See Table 1.1.)

Just as in two-dimensional plots, $z = f(x, y)$ cannot produce a surface in which two values of z would be present for a single doublet (x, y). A simple example of such a surface is a sphere $x^2+y^2+z^2 = a^2$. One can only produce half-spheres $z = \pm\sqrt{a^2 - x^2 - y^2}$. The remedy is to use a parametric plot. In fact, a parametric plot produces not only surfaces but also curves in three dimensions. Curves have a single parameter while surfaces involve two parameters. The two commands are

```
ParametricPlot3D[{f[t],g[t],h[t]},{t,a,b}]
ParametricPlot3D[{f[t,u],g[t,u],h[t,u]},{t,a,b},{u,c,d}]
```

where the first one produces a curve and the second one a surface in three dimensions. For example,

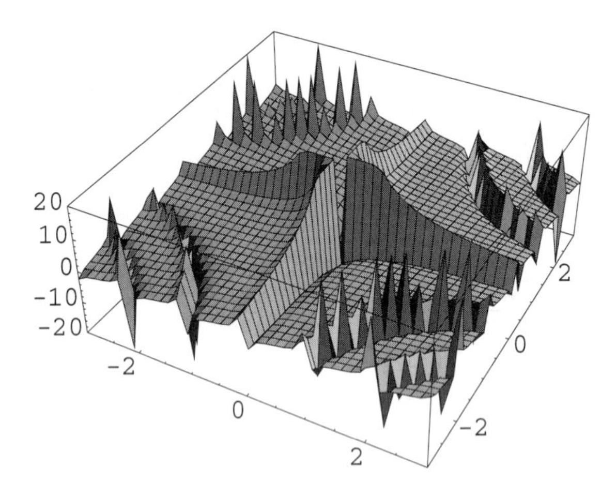

FIGURE 1.8. The three-dimensional plot of $\cot(tu)$.

```
In[2]:= ParametricPlot3D[{Sin[t], Cos[t], t/5},
        {t, 0, 6 Pi}]
```

produces the helix of Figure 1.9, and

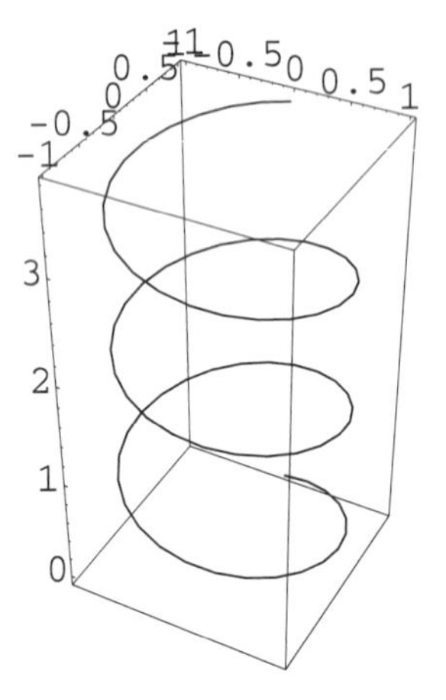

FIGURE 1.9. The helix is produced using **ParametricPlot3D** with a single parameter.

```
In[3]:= ParametricPlot3D[{Sin[t] Cos[u], Sin[t] Sin[u],
        Cos[t]},{t, 0, Pi},{u, 0, 2 Pi}]
```

produces the sphere of Figure 1.10.

In general, a function $f(\theta, \varphi)$ can be plotted parametrically as follows. When we plot $f(x, y)$, we treat the values of f as the third Cartesian

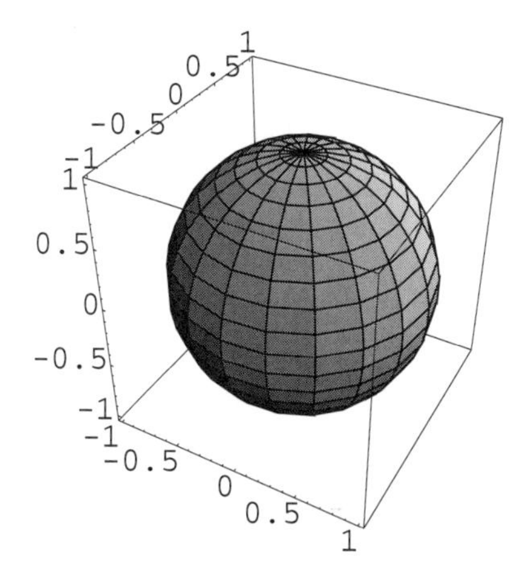

FIGURE 1.10. The sphere is produced using `ParametricPlot3D` with two parameters.

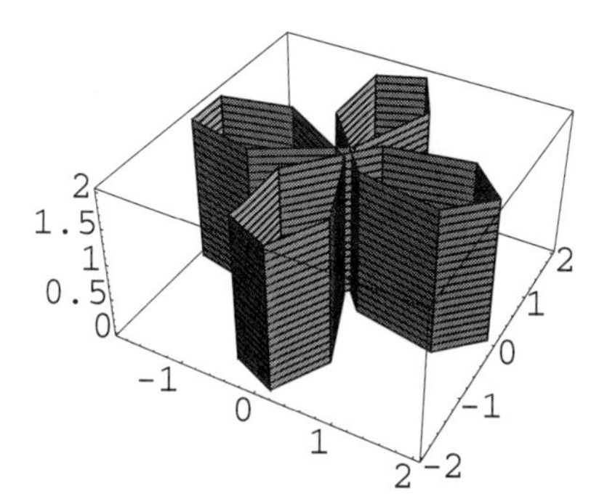

FIGURE 1.11. When only one of the axes is a function of one of the two parameters, a cylinder is drawn along that axis.

coordinate z. Similarly, for specific values of its arguments, $f(\theta, \varphi)$ can be thought of as the values of the third spherical coordinate r. Thus, writing $r = f(\theta, \varphi)$, we have

$$\begin{aligned} x(\theta, \varphi) &= r \sin\theta \cos\varphi = f(\theta, \varphi) \sin\theta \cos\varphi \\ y(\theta, \varphi) &= r \sin\theta \sin\varphi = f(\theta, \varphi) \sin\theta \sin\varphi \\ z(\theta, \varphi) &= r \cos\theta = f(\theta, \varphi) \cos\theta \end{aligned} \tag{1.3}$$

The three functions $x(\theta, \varphi)$, $y(\theta, \varphi)$, and $z(\theta, \varphi)$ can now be parametrically plotted to give a surface. In the special case of a sphere of unit radius, $f(\theta, \varphi) = 1$, and we get the three functions in `In[3]` above (with t replacing θ and u replacing φ).

2/3D	Option	Default value	Description
2D	`AspectRatio`	`1/GoldenRatio`	height-to-width ratio for the plot; `Automatic` sets it from absolute x- and y-coordinates
	`Axes`	`True`	whether to include axes; `False` draws no axes
	`AxesLabel`	`None`	whether to include axes
3D	`Boxed`	`True`	whether to draw a 3D box around a 3D plot
2D	`Frame`	`False`	whether to draw a frame around the plot
2D	`GridLines`	`None`	what gridlines to include; `Automatic` includes one gridline per major tick mark
2D	`PlotLabel`	`None`	expression used as plot label
3D	`PlotPoints`	`15`	number of points in each direction to sample the function
	`PlotRange`	`Automatic`	range of coordinates included in plot; `All` includes all points
2D	`Ticks`	`Automatic`	what tick marks to draw; `None` gives no tick marks
3D	`Shading`	`True`	whether to shade the surface

TABLE 1.1. Options for three-dimensional plots.

In a three-dimensional parametric plot if one of the two variables appears in only one of the axes, a "cylinder" is produced along that axis. For example,

```
In[3]:= ParametricPlot3D[{2 Cos[t] Cos[2t], 2 Sin[t]
        Cos[2t], u},{t, 0, Pi},{u, 0, 5}]
```

will yield the cylinder of Figure 1.11.

We have already used the **options** to alter the way *Mathematica* plots graphs by default. Table 1.1 lists some of the options for three-dimensional plots. Two-dimensional plots have fewer options, but they have a variety of styles.

1.7 Complex Numbers

For a long time it was thought that complex numbers were just toys invented and played with only by mathematicians. After all, no single quantity in the *real* world can be measured by an *imaginary* number, a number that lives only in the imagination of mathematicians.

However, things have changed enormously over the last couple of centuries. Continuing in the footsteps of Euler, who packaged the three most important numbers of mathematics in the formula $e^{i\pi} = -1$, Gauss took the complex numbers very seriously and, unlike his contemporaries who were reluctant to manipulate them as legitimate mathematical objects, treated them on an equal footing with the real numbers. The result was some fundamental conclusions concerning the number system itself, and the proof of the most important *fundamental theorem of algebra*. Cauchy, a French contemporary of Gauss, extended the concept of complex numbers to the notion of complex functions, and—almost single-handedly—developed the rich subject of complex analysis. By the end of the nineteenth century the subjects of complex algebra and calculus were firmly established in the entire mathematics community.

But no one in their wildest imagination could dream of a day when Nature itself would incorporate complex numbers in its most inner workings. In 1926 Erwin Schrödinger discovered that in the language of the world of the subatomic particles, complex numbers were the indispensable alphabets. The very first symbol in the Schrödinger equation is i. That is why the importance of complex numbers has not gone unnoticed in *Mathematica*.

We have already noted that *Mathematica* has the internal constant `I`, representing $\sqrt{-1}$, and have seen it find roots of polynomials as complex numbers. It is therefore not surprising to learn that *Mathematica* can manipulate complex numbers and functions. In fact, any symbol in *Mathematica* is treated as a complex quantity unless otherwise indicated.

MM Chapter 9 discusses complex numbers.

Some of the commands used in complex manipulations are

`Re[z], Im[z]`	real and imaginary parts of z
`Conjugate[z]`	complex conjugate of z
`Abs[z], Arg[z]`	absolute value $\|z\|$ and argument θ of z in $\|z\|e^{i\theta}$
`ComplexExpand[expr]`	expand *expr* assuming all variables are real

There is also `ComplexExpand[[expr],{x,y, ... }]`, which expands the `expr` assuming that $\{x, y, \ldots\}$ are complex.

We usually denote a complex number by z, with x and y its real and imaginary parts, respectively. As indicated above, *Mathematica* treats all variables as complex quantities, so x, y, and z are all such quantities. To see this, let us use `ComplexExpand`. The input

```
In[1]:= ComplexExpand[(x + I y)^2]
```

produces

Out[1]:= $x^2 + 2Ixy - y^2$

but

```
In[2]:= ComplexExpand[(x + I y)^2,{x,y}]
```

produces

```
Out[2]:=
```

$$-\operatorname{Im}[x]^2 - 2I\operatorname{Im}[x]\operatorname{Im}[y] + \operatorname{Im}[y]^2 + 2I\operatorname{Im}[x]\operatorname{Re}[x] - 2\operatorname{Im}[y]\operatorname{Re}[x]$$
$$+\operatorname{Re}[x]^2 - 2\operatorname{Im}[x]\operatorname{Re}[y] - 2I\operatorname{Im}[y]\operatorname{Re}[y] + 2I\operatorname{Re}[x]\operatorname{Re}[y] - \operatorname{Re}[y]^2$$

In the first case, x and y are treated as real numbers, eliminating the need for producing the real and imaginary parts of, e.g. x^2. In the second case, x and y are treated as complex numbers, each with its own real and imaginary parts.

1.7.1 An Example from Optics

Although no *single* measurable physical quantity corresponds to a complex number, *a pair* of physical quantities can be represented very naturally by a complex number. For example, a wave, which always consists of an amplitude and a phase, begs a representation by a complex number. Thus, the sinusoidal wave

$$A\cos(\mathbf{k}\cdot\mathbf{r} - \omega t + \phi) \tag{1.4}$$

representation of waves by complex numbers

can be thought of as the real part of the complex wave

$$Ae^{i(\mathbf{k}\cdot\mathbf{r}-\omega t+\phi)} \equiv Ze^{-i\omega t}$$

where Z is the *complex amplitude* $Ae^{i(\mathbf{k}\cdot\mathbf{r}+\phi)}$. In all discussions that follow in this section, $\mathbf{k}\cdot\mathbf{r}$ is treated as part of the phase and absorbed in ϕ.

The superposition of two waves $Z_1e^{-i\omega_1 t}$ and $Z_2e^{-i\omega_2 t}$ gives rise to a third "wave," which in general will have a time-dependent amplitude and no well-defined frequency. For example, the reader may verify that if $Z_1 = A_1e^{i\phi_1}$ and $Z_2 = A_2e^{i\phi_2}$, and if we could write the sum of the two waves as $\mathcal{A}e^{i(\phi-\omega t)}$, then

$$\mathcal{A}\cos(\phi - \omega t) = A_1\cos(\phi_1 - \omega_1 t) + A_2\cos(\phi_2 - \omega_2 t)$$
$$\mathcal{A}\sin(\phi - \omega t) = A_1\sin(\phi_1 - \omega_1 t) + A_2\sin(\phi_2 - \omega_2 t)$$

It turns out that $\mathcal{A}$, ϕ, and ω are all *time-dependent*. For example, by adding the squares of the two equations above, we obtain

$$\mathcal{A} = \sqrt{A_1^2 + A_2^2 + 2A_1A_2\cos[(\phi_1 - \phi_2 + (\omega_1 - \omega_2)t]}$$

However, if the two frequencies are equal, with $\omega = \omega_1 = \omega_2$, then the entire time dependence of the superposition of the two waves can be described by this common frequency, and the amplitude will be time-independent.

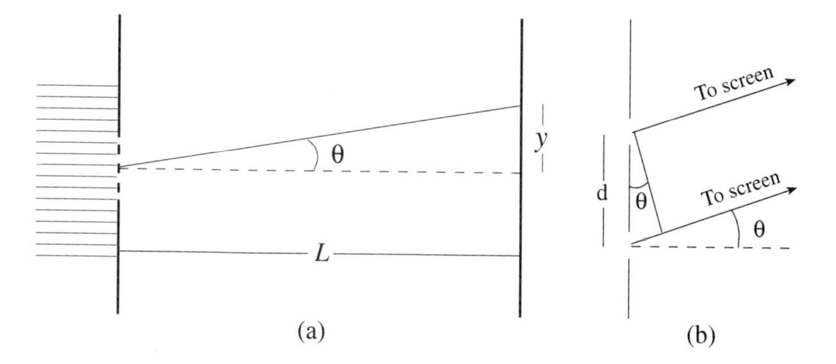

FIGURE 1.12. (a) An N-slit arrangement, showing the viewing angle θ. (b) Two adjacent slits and the rays that meet at the screen far away.

The problem of the superposition of a set of waves with the same frequency reduces to adding their complex amplitudes:

$$\sum_{j=1}^{N} Z_j e^{-i\omega t} = \left(\sum_{j=1}^{N} A_j e^{i\phi_j} \right) e^{-i\omega t} \equiv \mathcal{A} e^{-i\omega t}$$

In general, the sum in parentheses cannot be calculated in closed form. However, a very special (but useful) case lends itself to an analytic solution. In an N-slit interference (Figure 1.12), all A_j's are equal, and $\phi_j = j\phi$, where ϕ is the phase difference between the waves of consecutive slits acquired due to the difference in path length as the waves move from their sources to the location of interference. We now use *Mathematica* to find the intensity of an N-slit interference pattern.

N-slit interference

Recall from introductory physics that the intensity is (proportional to) $\frac{1}{2}|amp|^2$, which for a single slit of amplitude A becomes $\frac{1}{2}|A|^2$. Calling this intensity I_1, the N-slit intensity I_N can be written as

$$I_N = \tfrac{1}{2}|\mathcal{A}|^2 = \tfrac{1}{2}|A|^2 \left| \sum_{j=1}^{N-1} e^{ij\phi} \right|^2 = I_1 \left| \sum_{j=1}^{N-1} e^{ij\phi} \right|^2$$

where the sum extends to $N-1$ because there are $N-1$ phase *differences* between N slits. Our job is to calculate the sum $\sum_{j=1}^{N-1} e^{ij\phi}$. Typing in

```
In[1]:= amp[N_,phi_]:=Sum[E^(I k phi),{k,0,N-1}]
```

yields

```
Out[1]:=
```

$$\frac{-1 + E^{IN\phi}}{-1 + E^{I\phi}}$$

From here on we have to get control of the manipulations because otherwise *Mathematica* assumes that all quantities are complex. So, the first thing we want to do is to write the numerator in terms of trigonometric functions. This is accomplished by

```
In[2]:= num1[N_,phi_]=ComplexExpand[Numerator[amp[N,phi]]]
```

Out[2]:=

$$1 - \mathrm{Cos}[N\phi] + I\,\mathrm{Sin}[N\phi]$$

Since we are interested in intensity, we need the square of the absolute value of the numerator:

```
In[3]:= numInt[N_,phi_]= Simplify[ComplexExpand[num1[N,phi]
        Conjugate[num1[N,phi]]]]
```

Out[3]:=

$$2 - 2\,\mathrm{Cos}[N\phi]$$

We do the same thing with the denominator:

```
In[4]:= den1[N_,phi_]=ComplexExpand[Denominator[amp[N,phi]]]
```

Out[4]:=

$$1 - \mathrm{Cos}[\phi] + I\,\mathrm{Sin}[\phi]$$

```
In[5]:= denInt[N_,phi_]= Simplify[ComplexExpand[den1[N,phi]
        Conjugate[den1[N,phi]]]]
```

Out[5]:=

$$2 - 2\,\mathrm{Cos}[\phi]$$

Setting $I_1 = 1$, we finally obtain

```
In[6]:= intensity[N_,phi_]=Simplify[numInt[N,phi]
        /denInt[N,phi]]
```

Out[5]:=

$$\frac{1 - \mathrm{Cos}[N\phi]}{1 - \mathrm{Cos}[\phi]}$$

This is sometimes written as

$$I_N = I_1 \frac{1 - \cos(N\phi)}{1 - \cos\phi} = I_1 \left[\frac{\sin(N\phi/2)}{\sin(\phi/2)}\right]^2, \quad \phi = \left(\frac{2\pi}{\lambda}\right) d \sin\theta \tag{1.5}$$

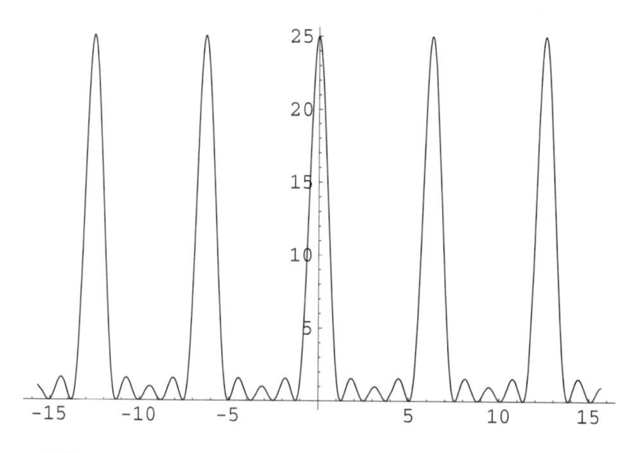

FIGURE 1.13. The intensity of a five-slit arrangement.

where the single-slit intensity I_1 has been reintroduced, and ϕ has been written explicitly in terms of the viewing angle θ. To see where this expression for ϕ comes from, refer to Figure 1.12(b) and note that the path difference between the two adjacent rays is $d \sin\theta$. Now recall from Equation (1.4) that the phase difference *due to the path difference*—this is the only phase difference considered, because the ϕ of Equation (1.4) is assumed to be zero—is simply $k = 2\pi/\lambda$ times the path difference.

Figure 1.13, produced by the command

```
In[7]:= Plot[intensity[5,phi],{phi,-5 Pi,5 Pi},
        PlotRange->All]
```

shows the intensity of a five-slit arrangement. Notice that there are some *primary* maxima, in between which we find four minima and three secondary maxima. The fact that there are $N-1$ minima and $N-2$ secondary maxima between any two primary maxima of an N-slit arrangement is depicted in Figure 1.14, where the intensity of a 3-slit, a 7-slit, and a 10-slit arrangement is plotted as a function of ϕ. Note that the 3-slit intensity shows 2 minima and 1 secondary maxima, the 7-slit 6 minima and 5 secondary maxima, and the 10-slit 9 minima and 8 secondary maxima. We now want to use *Mathematica* to investigate the nature of these extrema.

The locations of extrema of a function are the roots of the derivative of that function. So, let us differentiate the intensity

```
In[1]:= derInt[N_,phi_]=Together[D[intensity[N,phi],phi]]
```

```
Out[1]:=
```

$$\frac{-\operatorname{Sin}[\phi] + \operatorname{Cos}[N\phi]\operatorname{Sin}[\phi] + N\operatorname{Sin}[N\phi] - N\operatorname{Cos}[\phi]\operatorname{Sin}[N\phi]}{(1-\operatorname{Cos}[\phi])^2}$$

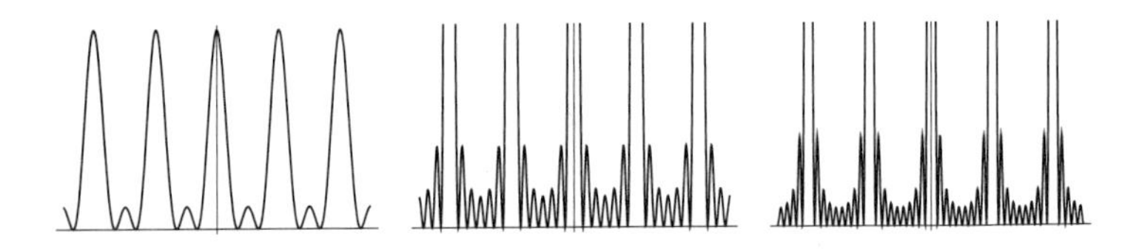

FIGURE 1.14. The intensity of a three-slit (left), a seven-slit (middle), and a ten-slit arrangement plotted as a function of ϕ.

Without `Together`, *Mathematica* would separate the answer into two fractions. Now find the roots of the derivative:

```
In[2]:= rts[N_,r_]:=FindRoot[derInt[N,phi]==0,{phi,r}];
        extremum[N_,r_]:=phi /. rts[N,r]
```

This little routine finds the roots of the derivative near $\phi = r$ and calls them `extremum[N,r]`. To feed a good r into this routine, it would be helpful if we plotted the derivative. We now look at the special case of 8 slits in some detail. The reader may wish to try other values of N.

For $N = 8$, the plot of the derivative looks like Figure 1.15. Note that there are 15 roots altogether. The roots at the two extremes of the figure occur at $\phi = 0$ and $\phi = 2\pi$ and correspond to the two primary maxima there. The roots with a positive slope—there are seven of these—correspond to the minima between two primary maxima.[4] The remaining roots with negative slopes—there are six of these—correspond to the secondary maxima between two primary maxima.

The graph gives us only the rough estimate of the locations of the roots, and we can use these estimates to calculate the more accurate locations. For example, for $N = 8$, the first minimum appears to occur around $r = 0.8$. Thus, we type in

```
In[3]:= min8[1]=extremum[8,0.8]
```

and obtain

```
Out[3]:= 0.785398
```

Typing in `min8[2]=extremum[8,1.6]`, etc., we find the location of the seven minima of the intensity. Then we use the command

```
In[4]:= tabMin=Table[min8[j],{j,1,7}]
```

to produce the following list:

```
Out[4]:=
```

[4]Recall that an extremum where the second derivative—i.e., the slope of the graph of the derivative—is positive (negative) is a minimum (maximum).

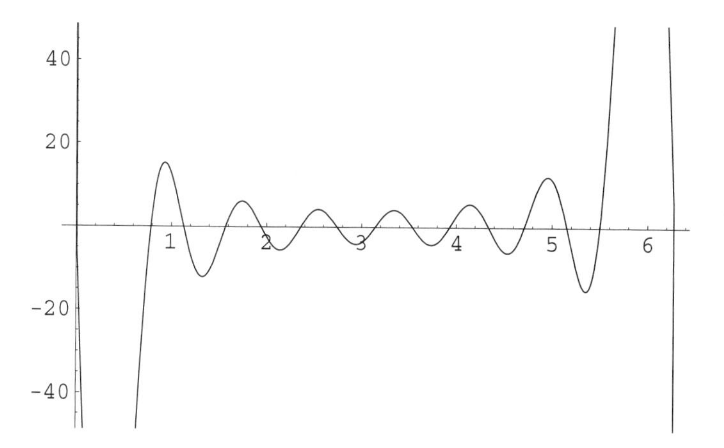

FIGURE 1.15. The derivative of the intensity of an eight-slit arrangement.

$$\{0.785398, 1.5708, 2.35619, 3.14159, 3.92699, 4.71239, 5.49779\}$$

Similarly, we can find the secondary maxima. In Figure 1.15 these correspond to the horizontal intercepts with negative slopes. The first one occurs about $r = 1.2$. So, we type in

```
In[5]:= secMax8[1]=extremum[8,1.2]
```

and obtain

```
Out[5]:= 1.12939
```

As before, we create a list of the locations of the secondary maxima—after evaluating `secMax8[2]` through `secMax8[6]`—by using the command

```
In[6]:= tabMax=Table[secMax8[j],{j,1,6}]
```

which yields

```
Out[6]:=
```

$$\{1.12939, 1.94218, 2.74258, 3.54061, 4.341, 5.15379\}$$

It is instructive to calculate the intensities at the minima (which we expect to be zero) and maxima. This is easily done by substitution in the intensity formula. However, because of the denominator, we expect indeterminate results. Therefore, the command `Limit` has to be used. So, we type in

```
In[7]:= tabIntMin=Table[Limit[intensity[8,phi],
        phi->min8[j]],{j,1,7}]
```

and get

Out[7]:=

$$\{0, 0, 0, 0, 0, 0, 0\}$$

as we should.

For the intensity of the secondary maxima, we type in

```
In[8]:= tabIntSecMax=Table[Limit[intensity[8,phi],
        phi->secMax8[j]],{j,1,6}]
```

and get

Out[8]:=

$$\{3.36082, 1.45681, 1.04022, 1.04022, 1.45681, 3.36082\}$$

For primary maxima, we need to evaluate the intensity at $\phi = 2m\pi$ for the integer m. These will all give the same result by Equation (1.5). Mere substitution, however, will give an indeterminate result. Therefore, we have to take limits. As a convenient shortcut, we define the following command, which is good not only for the primary maxima, but for all extrema:

```
In[9]:= IntAtExtrema[N_,r_]:=Limit[intensity[N,phi],
        phi->extremum[N,r]]
```

Let us use this to find the intensity of primary maxima for different number of slits. We use $r = 2\pi \approx 6.2832$ for $N = 2$ to $N = 10$ in the following command to make a table:

```
In[10]:= PrimMax=Table[{j,IntAtExtrema[j,6.2832]},
         {j,2,10}]
```

Executing this command yields

Out[10]:=

$$\{\{2, 4.\}, \{3, 9.\}, \{4, 16.\}, \{5, 25.\}, \{6, 36.\}, \{7, 49.\}, \{8, 64.\},$$
$$\{9, 81.\}, \{10, 100.\}\}$$

It is clear from the above list that the intensity of the primary maxima should be equal (actually proportional) to the square of the number of slits. We can show this analytically by evaluating the limit of the intensity at $\phi = 0$. In fact, typing

```
In[11]:= Limit[intensity[N,phi],phi->0]
```

yields N^2 as the output. Recalling that the single-slit intensity was taken to be 1, we obtain the well-known result from optics that $I_N^{\text{prim max}} = N^2 I_1$.

The preceding discussion of maxima and minima assumes that the single-slit intensity I_1 is independent of the viewing angle θ. This assumption is

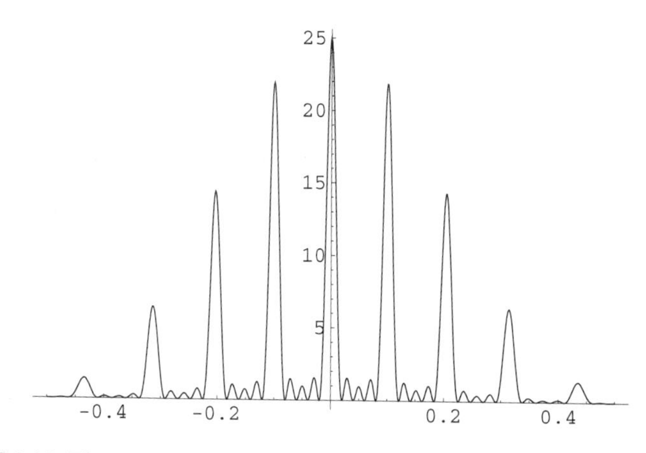

FIGURE 1.16. The interference pattern of a five-slit arrangement with diffraction.

not strictly correct, because a slit can produce an interference pattern of its own. The Huygens' principle in wave optics tells us that the wave emerging from a single slit is to be considered as a collection of point sources capable of producing their own interference pattern. Starting with Equation (1.5), and assuming that the single slit of width a is a collection of N narrow slits separated by a/N, it is not hard (see Problem 1.28 for details) to show that, in the limit of $N \to \infty$, one obtains the single-slit **diffraction** formula:

$$I_1 = I_0 \left[\frac{\sin(\beta/2)}{\beta/2}\right]^2, \qquad \beta = \frac{2\pi}{\lambda} a \sin\theta \tag{1.6}$$

where I_0 is the single-slit intensity at the viewing angle $\theta = 0$. Combining this with (1.5), we obtain

$$I_N = I_0 \left[\frac{\sin(\beta/2)}{\beta/2}\right]^2 \left[\frac{\sin(N\phi/2)}{\sin(\phi/2)}\right]^2, \quad \beta = \frac{2\pi}{\lambda} a \sin\theta, \quad \phi = \frac{2\pi}{\lambda} d \sin\theta \tag{1.7}$$

where a is the width of each of the N slits and d is their separation.

To see the effect of diffraction on the N-slit interference, we look at the plot of the I_N of (1.7) as a function of y, the viewing location in Figure 1.12. To this end, we type in

```
beta[y_,a_,lambda_,L_]:=(2 Pi a/lambda) y/Sqrt[y^2+L^2];
  phi[y_,d_,lambda_,L_]:=(2 Pi d/lambda) y/Sqrt[y^2+L^2];
  diff[y_,a_,lambda_,L_]:=(Sin[beta[y,a,lambda,L]/2]
        /(beta[y,a,lambda,L]/2))^2;
  interf[y_,d_,lambda_,n_,L_]:=(Sin[n phi[y,d,lambda,L]/2]
        /(Sin[phi[y,a,lambda,L]/2]))^2;
```

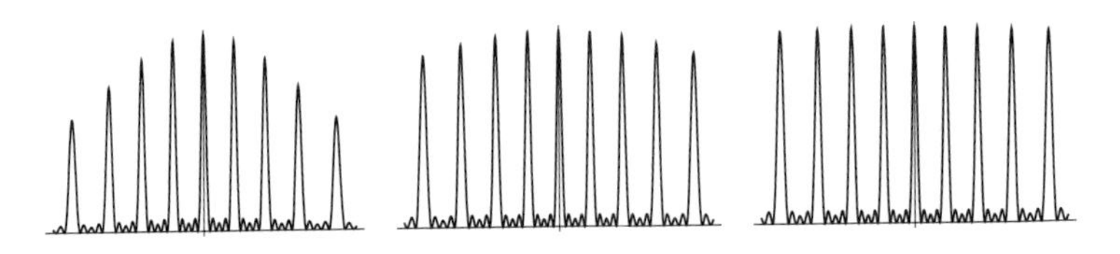

FIGURE 1.17. The interference pattern of a five-slit arrangement with diffraction with $d/a = 10$ on the left, $d/a = 20$ in the middle, and $d/a = 50$ on the right.

```
totIntensity[y_,d_,a_,lambda_,n_,L_]:=diff[y,a,lambda,L]
      interf[y,d,lambda,n,L]
```

Then the input

```
In[2]:= Plot[totIntensity[y,0.05,0.01,0.005,5,1],
        {y,-0.5,0.5},PlotRange->All]
```

will produce the plot shown in Figure 1.16. Notice that the intensity of the primary maxima is sharply reduced away from the central maximum. This effect is due to the diffraction caused by each slit; an effect that can be reduced by taking d, the slit spacing, to be much larger than a, the width of each slit. We can exhibit this by plotting `totIntensity` for—say three—different ratios d/a, calling these plots `pl1`, `pl2`, and `pl3`, and using

use of **Show** and **GraphicsArray**

```
Show[GraphicsArray[{pl1,pl2,pl3}]]
```

The result is exhibited in Figure 1.17. Note that as the ratio d/a increases, the effect of diffraction decreases.

1.8 Animation

One of the most powerful features of *Mathematica* is the ease with which animations can be achieved. When a physical quantity is a function of time in addition to other variables (such as spatial coordinates), animation becomes a great visual aid in understanding the behavior of the quantity. In textbooks the effect of animation is sometimes depicted as a series of still images (snapshots) of the quantity in question at different times. *Mathematica* can produce these snapshots in sufficient quantities so that their rendering in succession will give the illusion of motion just as the successive exposure of the frames of a movie creates the illusion of motion. The commands used in such situations are

`<<Graphics‘Animation‘`	load the Animation package
`Animate[plot,{t,a,b,dt}]`	animate `plot` with t running from a to b in steps of dt

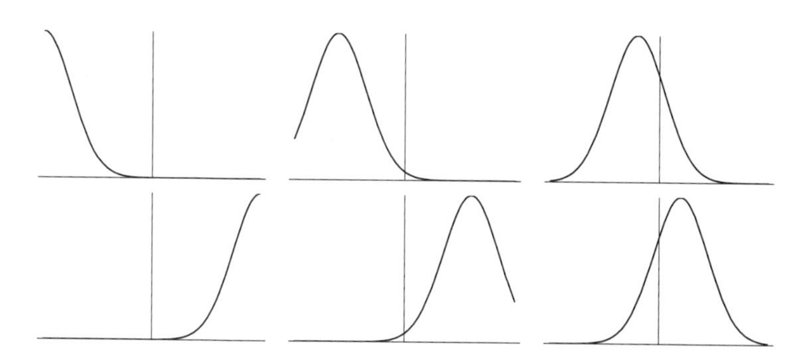

FIGURE 1.18. Some of the snapshots that *Mathematica* created in the animation of a Gaussian pulse. The upper-left frame corresponds to $t = 0$, with t increasing clockwise, ending at $t = 5$ at the lower left.

In some platforms `<<Graphics`Animation`` is not necessary. The argument of **`Animate`** is one of the plot commands of *Mathematica*, which acts on a function, one of whose independent variables is t (usually time). The output of this command is a series of graphics cells, each slightly different from the previous one. To see the animation, one has to select the graphics cells and choose **Animate Selected Graphics** from the **Cell** menu. The keyboard shortcut for this is `Command+y` in Macintosh and `Control+y` in Windows. Let us look at some examples.

How to play an animation

First, consider a Gaussian-shaped pulse moving in the positive x-direction. The reader recalls that a traveling pulse having speed v and a shape described by the function $f(x)$ is given by $f(x \pm vt)$, where the plus and the minus represent, respectively, a left-moving and a right-moving pulse. For simplicity, we assume that the speed is unity, and that the pulse starts out at $x = -2$. Then, the Gaussian pulse will have the form $e^{-(x+2-t)^2}$, and the command

```
In[1]:= Animate[Plot[Exp[-(x+2-t)^2],{x,-2,2},
        PlotRange->{0,1}],{t,0,5,0.2}]
```

will create 26 plots starting with $t = 0$ and ending with $t = 5$, separated by intervals of 0.2 units. Selecting these plots and pressing the combination `Command+y` keys in Macintosh (`Control+y` keys in Windows) will play the movement of the Gaussian pulse from left to right and loop it back to the beginning after reaching the end. Figure 1.18 shows some of the snapshots created by *Mathematica* in the animation of our Gaussian pulse.

Next, we consider the formation of beats when two sinusoidal waves of slightly different frequencies interfere. This time let's make the waves move to the left, but keep their speed the same.[5] Thus, we type in

[5]The wave number (coefficient of x) and the angular frequency (coefficient of t) of the two waves are different, but their ratios are the same.

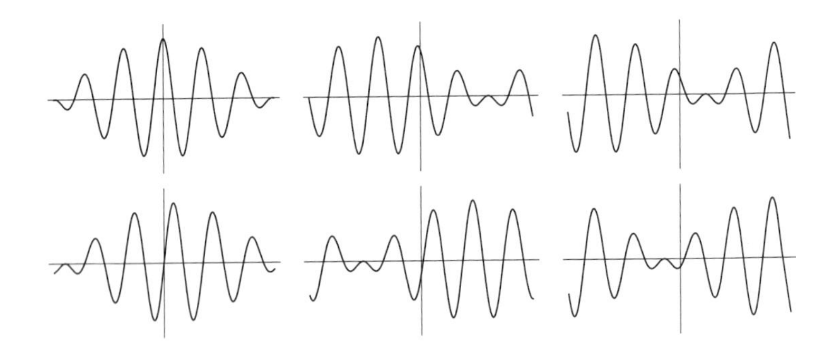

FIGURE 1.19. Some of the snapshots created by *Mathematica* in the animation of a beat. The upper-left frame corresponds to $t = 0$, with t increasing clockwise, ending at $t = 30$ at the lower left.

```
In[2]:= f[x_,t_]:=Cos[x+t]+Cos[1.2 x+1.2t];
        Animate[Plot[f[x,t],{x,-16,16},
        PlotRange->{-2.5,2.5}],{t,0,30,1}]
```

following which *Mathematica* creates 31 plots starting at $t = 0$ and separated consecutively by one unit of t. Some of these plots are shown in Figure 1.19. Note the specification of `PlotRange` in this animation. If we do not specify the plot range, and the range of the plots are different at different t's, then *Mathematica* automatically *rescales* the plots, causing the plots to have essentially different units for axes at different times.

Our third example is a two-dimensional wave that travels diagonally in the xy-plane. The input is

```
In[3]:= Animate[Plot3D[Cos[x+y-t],{x,0,2Pi},{y,0,2Pi}],
        {t,0,2Pi}]
```

By choosing the range of x, y, and t to be 0 to 2π, the animation repeats itself smoothly and continuously. Figure 1.20 shows some of the snapshots created by *Mathematica* in the animation of this two-dimensional wave. The input

```
In[4]:= Animate[Plot3D[Cos[x-t],{x,0,2Pi},{y,0,2Pi}],
        {t,0,2Pi}]
```

the option **ViewPoint** for a 3D plot

creates a surface wave traveling along the x-axis. It resembles the waving of a flag viewed from *Mathematica*'s default viewing angle. To change this angle, one changes the values of `Plot3D` option `ViewPoint`. These values give the coordinates of the point in space from which to look at the surface. Changing the viewpoint as follows:

```
In[5]:= Animate[Plot3D[Cos[x-t],{x,0,2Pi},{y,0,2Pi}],
        {t,0,2Pi},ViewPoint->{0.4,0.5,2}]
```

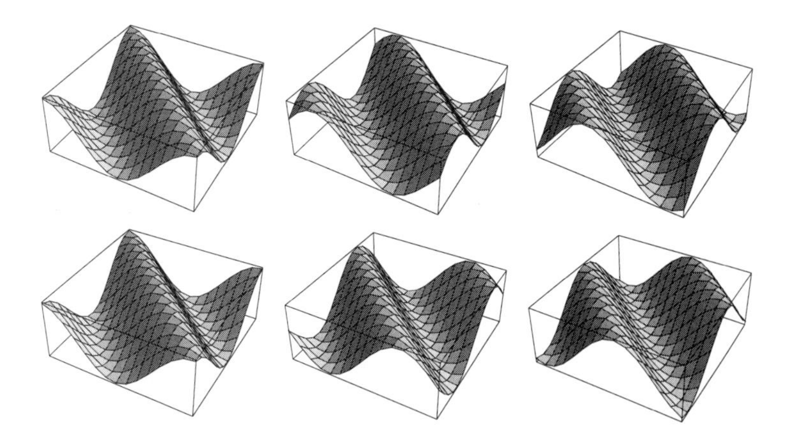

FIGURE 1.20. Some of the snapshots created by *Mathematica* in the animation of a two-dimensional wave. The upper-left frame corresponds to $t = 0$, with t increasing clockwise, ending at $t = 2\pi$ at the lower left.

the waving sheet appears on the screen as if looked at from the top.

Our last example has to do with the time-dependent solutions of the one-dimensional Schrödinger equation in an infinite potential well of width L. It is well known from introductory physics that the stationary solutions (energy eigenstates) are of the form

$$\psi_n(x) = \sqrt{\frac{2}{L}} \sin\left(\frac{n\pi x}{L}\right), \qquad n = 1, 2, 3, \ldots$$

It is also known that the time-dependent solutions are obtained from the stationary solutions by a superposition of a number of such states each multiplied by $e^{-i\omega_n t}$, with $\omega_n = E_n/\hbar$, where E_n is the energy of the nth state. More specifically, a time-dependent solution of this particular Schrödinger equation can be written as

$$\Psi(x,t) = \sum_{k=1}^{N} a_k \sin\left(\frac{k\pi x}{L}\right) e^{-iE_k t/\hbar}, \qquad E_k = \frac{\hbar^2\pi^2}{2mL^2} k^2 \tag{1.8}$$

where $\sum_{k=1}^{N} |a_k|^2 = 1$. If all a_k are zero except for one, say a_n, then

$$\Psi(x,t) = a_n \sin\left(\frac{n\pi x}{L}\right) e^{-iE_n t/\hbar} \Rightarrow |\Psi(x,t)|^2 = |a_n|^2 \sin^2\left(\frac{n\pi x}{L}\right)$$

implying a time-independent probability density.[6] That is why these states are called *stationary*.

[6]For those unfamiliar with quantum theory, the square of the absolute value of the solution of the Schrödinger equation is the probability density for the (subatomic) particle involved.

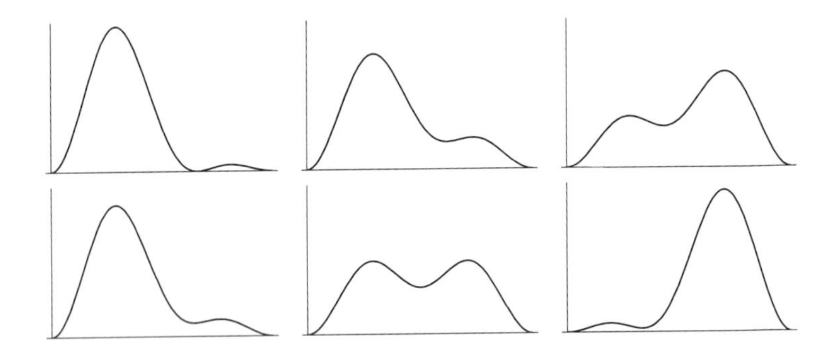

FIGURE 1.21. Some of the snapshots created by *Mathematica* in the animation of an infinite quantum well. The upper-left frame corresponds to $t = 0$, with t increasing clockwise, ending at $t = 2$ at the lower left.

A time-dependent probability, therefore, requires a superposition of at least two different stationary states. So, let us type in

```
In[6]:= psi[x_,t_,n_]:= Sum[a[j] Sin[j Pi x]
        Exp[-I Pi j^2t],{j,1,n-1}]+Sqrt[1-Sum[a[m]^2,
        {m,1,n-1}]] Sin[n Pi x]Exp[-I Pi n^2 t];
        p[x_,t_,n_]:= ComplexExpand[psi[x,t,n]
        Conjugate[psi[x,t,n]]]
```

Here

$$\Psi(x,t) = \sum_{j=1}^{n-1} a_j \sin(j\pi x) e^{-i\pi j^2 t} + \sqrt{1 - \sum_{m=1}^{n-1} a_m^2}\, \sin(n\pi x) e^{-i\pi n^2 t}$$

is a superposition of n solutions, whose probability density is also defined in `In[6]`. All constants such as the width of the potential, the particle's mass, etc. are set equal to 1. For $n = 2$ and $a_1 = 1/\sqrt{2}$, we type in

```
In[7]:= a[1]:=1/Sqrt[2]; Animate[Plot[p[x,t,3],{x,0,1},
        Axes->False,PlotRange->{0,1.6}],{t,0,2Pi/3,0.05}]
```

and obtain 42 graphics cells whose animation shows the probability density sloshing back and forth between the left and right halves of the potential well. The specification of the `PlotRange` is necessary to avoid the rescaling of the plots at different times. Figure 1.21 shows some of the snapshots created by *Mathematica* in the animation of this infinite quantum well. The period of the oscillation between the left and the right halves of the well turns out to be $2\pi/3$, as the reader may verify by evaluating $|\Psi(x,t)|^2$.

1.9 Input and Output Control

In this final section of our overview we want to say a few words about how input and output of *Mathematica* are controlled. Let us first consider the output.

We have already seen how to control the numerical output by specifying the number of decimal places using `N[ ]`. If you end the *Mathematica* input expression with `// N`, the expression is evaluated numerically and the output is given with the default number of significant figures. For example, the input `Pi // N` gives 3.14159.

A similar command exists for symbolic calculations as well. This command is especially useful if the output is extremely long. For instance, typing in

```
In[1]:= Expand[(2 a + 3 b - 4)^10]
```

gives a result 16 lines long. On the other hand,

```
In[2]:= Expand[(2 a + 3 b - 4)^10] // Short
```

yields

Out[2]:= $1048576 - 5242880a + << 91 >> +393660ab^9 + 59049b^{10}$

indicating that the output has ignored 91 terms of the result and printed only the given 4 terms. We can control the number of output lines by using the command `Short[`*expr*`,n]`, which gives the outline of the answer in n lines. It also gives the number of terms it has suppressed. For example,

Short[expr,n] explained

```
In[3]:= Short[Expand[(2 a + 3 b - 4)^10],3]
```

gives the answer in three lines, including the symbol `<<79>>`, which indicates that 79 terms of the answer have been suppressed.

meaning of <<>> in output

If you want the output to be completely suppressed, you can use a semicolon after the command. Furthermore, you can input several commands at the same time by separating them with semicolons. For example,

use of semicolon to suppress output

```
In[4]:= f[x_]=x^2; g[t_]=Sin[t]^3;
```

does not reproduce the input—as *Mathematica* normally does. However, it stores the two functions defined and remembers them later. This can be seen by typing in

```
In[5]:= f[3] + g[Pi/4]
```

which yields $9 + 1/2\sqrt{2}$.

Starting with *Mathematica* 3, you can input many expressions by simply clicking on the symbols provided in various **palettes**. There are seven built-

using **palettes** for inputting expressions

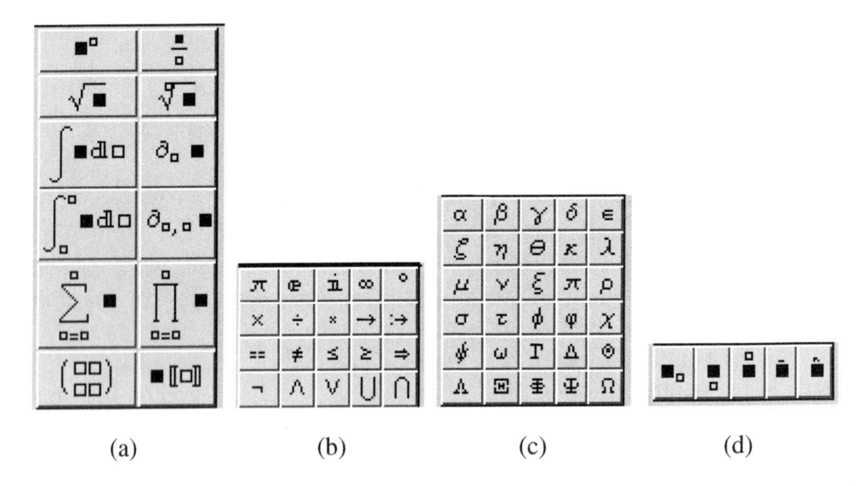

FIGURE 1.22. Different "subpalettes" of the `BasicInput` palette. The palettes of (a) mathematical operations, (b) widely used mathematical symbols, (c) Greek alphabet, and (d) subscripts, etc.

in palettes:[7] `Algebraic Manipulation`, `Basic Calculations`, `Basic Input`, `Basic Typesetting`, `Complete Characters`, `International Characters`, and `Notebook Launcher`. All these can be accessed from the **File** pop-up menu. The first three are the most relevant to our purposes.

A palette is a collection of **buttons**.

A palette is a collection of **buttons** arranged in rows and columns. Each button produces its symbol in the notebook upon being clicked. You can put several palettes next to each other. For instance, the `BasicInput` palette consists of four "subpalettes," which are shown separately in Figure 1.22. Although the buttons are self-explanatory (for instance, $\partial_{t,u}$ ■ stands for the mixed derivative with respect to t and u of an expression in the black box), a few remarks are in order.

The black squares are in general the first entries, i.e., when you click a button with a black square, what you type first will end up in that square. To type in other placeholders, you need to click on them. You can use the buttons *after* you have typed the main expression (what goes in the black square). Simply select what is intended for the black square, and *then* click on the button. Suppose you want to input `Sin[x y]+Cos[x]^3`. You can type in `Sin[x y]+Cos[x]`, select `Cos[x]`, and click on the first button of the palette shown in Figure 1.22(a). Now enter 3 for the exponent.

The second palette, depicted in Figure 1.22(b), contains symbols that are used in standard mathematics or in *Mathematica* (or both). The first entry is π, the famous number in mathematics that is also represented in

[7] *Mathematica* allows you to create palettes of your own. However, a discussion of such details would take us too far afield.

Mathematica by `Pi`. You cannot use π as a (constant or variable) Greek letter. The second entry is the base of the natural logarithm; and the third entry is $\sqrt{-1}$.

The third palette [Figure 1.22(c)] is the Greek alphabet, mostly lower case, but some capital letters are also included. Note that π is also there, and it has the same meaning as the first entry of the previous palette. Typing `N[π]` and hitting **Return** will produce 3.14159.

The last palette [Figure 1.22(d)] shows some of the so-called *modifier* buttons. These are symbols—such as subscripts and *underscripts*—added to variables to make them more in tune with the common usage in mathematics.

The other palettes are more specialized but also easy to use. We shall have many occasions to use them in the remainder of this book. It is a good idea to get into the habit of using these palettes, because they save a lot of typing, and you don't have to memorize all the commands.

Because of typographical limitations, we continue to use input commands in their text format. However, use of palettes saves a lot of typing, and the reader is well advised to make frequent use of them.

1.10 Problems

Problem 1.1. Consider the integers 137174210 and 1111111111. Put a decimal point at the end of one of them, then take the ratio of the first to the second. Using `N`, ask *Mathematica* to give you the result to 50 decimal places. Now remove the inserted decimal point and do the same thing. You should get a surprisingly interesting result!

Problem 1.2. Find the numerical value of i^i, where $i = \sqrt{-1}$. Compare your result with the numerical value of $e^{-\pi/2}$.

Problem 1.3. Find the factorial of 100. Of 300. Of 500. Now try the factorial of 4.5 and 3.2, or any other noninteger number.

Problem 1.4. Factor $x^{10} - 1$. Manipulate the result to get the original expression back.

Problem 1.5. Using `FactorInteger`, find the prime factors of 9699690 and 9765625.

Problem 1.6. Have *Mathematica* evaluate $(a+b)^{100}$ for you. Then tell it to find the coefficient of $a^{50}b^{50}$. Check to see if this coefficient is what you expect.

Problem 1.7. Start with the expression $x^4 + y^4$. Substitute $\cos z$ for x and $\sin z$ for y. Then ask *Mathematica* to *reduce* the result for you.

Problem 1.8. Instruct *Mathematica* to find the indefinite integral of $1/(1+x^2)$. Now change the power of x to 3, 4, etc., and see what the result will be.

Problem 1.9. Instruct *Mathematica* to find the indefinite integral of $1/\sin x$. Now differentiate the answer and `TrigReduce` it. Is the end result the same as the original integrand?

Problem 1.10. Define a function $g(x)$ as

$$g(x) = \int_x^{x^3} \sin(xt^2)dt$$

Then ask *Mathematica* to find its derivative. Use the **Help** menu to find out about functions introduced in the answer.

Problem 1.11. Define a function $g(x)$ as

$$g(x) = \int_{\sin x}^{\cos x} e^{x^2 t^3} dt$$

(a) Ask *Mathematica* to find its derivative. Use the **Help** menu to find out about functions introduced in the answer.
(b) Find the numerical value of $g'(3)$. See if you can find the source of the imaginary part of $g'(3)$. After all, a real function such as the integral defining $g(x)$ cannot have a complex derivative!

Problem 1.12. Find the roots of the following polynomials in terms of radicals:

$$\begin{aligned}
&(a) \quad x^3 - 3x^2 + 5x - 2 \\
&(b) \quad 3x^3 - 2x^2 + 7x - 9 \\
&(c) \quad x^4 - 5x^3 - 4x^2 - 6x + 3 \\
&(d) \quad -6x^4 + 3x^3 - 2x^2 - 7x + 11 \\
&(e) \quad x^5 - x^4 + 5x^3 - 7x^2 - 3x + 13
\end{aligned}$$

Now find the *numerical* values of the roots of the same polynomials.

Problem 1.13. Find at least three different values of x for which e^x and $\tan x$ are equal.

Problem 1.14. Find all values of x for which $e^x = 4x$.

Problem 1.15. Find all values of x for which $e^{2x} = 2 - x$.

Problem 1.16. Are there any angles for which the sine of the angle is equal to the cosine of the square of the angle? If so, find a few such angles.

Problem 1.17. Define functions $f1(x,n)$, $f2(x,n)$, $f3(x,n)$, and $f4(x,n)$ to be the sum of the first n terms of the Maclaurin series of e^x, $\sin x$, $\cos x$, and $\ln(1+x)$, respectively. For some values of x and n, compare the functions with their finite-sum representations.

Problem 1.18. In this problem, we want to find a series representation of π for *Mathematica*.
(a) In *Mathematica* define the function $a(x,n)$ to be the `Evaluation` of the nth derivative of $\arcsin(x)$.
(b) Using (a), define $pi(y,k)$ to be the sum of the first k terms of the Maclaurin series for $\arcsin(y)$.
(c) Substituting 0.5 for y, find a series (really a polynomial of degree k) for $\pi/6$.
(d) Compare π and the series by evaluating the differences $\pi - 6 * pi(0.5, 10)$, $\pi - 6 * pi(0.5, 20)$, $\pi - 6 * pi(0.5, 40)$, and $\pi - 6 * pi(0.5, 100)$.

Problem 1.19. Plot the difference between e^x and its Maclaurin series expansion up to x^6 for x between 0 and 5. Use `PlotRange->All` as an option of the plot. Do the same with the Maclaurin series expansion up to x^{10}.

Problem 1.20. Define a function

$$g(x) = \int_0^{\pi/2} \sqrt{1 - x^2 \sin^2 t}\, dt$$

Then plot g in the interval $0 < x < 0.5$.

Problem 1.21. Define a function

$$g(x,y) = \int_0^y \sin(xt^2)\, dt$$

Then make a three-dimensional plot of g in the interval $0 < x < 2$, $0 < y < 2\pi$.

Problem 1.22. For points in the xy-plane, find the electrostatic potential of four point charges equal in magnitude with two positive charges located at $(1,0)$ and $(0,1)$, and two negative charges located at $(-1,0)$ and $(0,-1)$. Now make a three-dimensional plot of this potential for the range $-2 < x < 2$, $-2 < y < 2$. Make sure you have enough `PlotPoints` to render the plot smooth.

See [Hass 00, pp. 27–28], for electrostatic potential of point charges.

Problem 1.23. Make a contour and a density plot of the potential of the previous problem.

Problem 1.24. Spherical harmonics $Y_{lm}(\theta, \varphi)$, where l is a nonnegative integer and m takes integer values between $-l$ and $+l$, occur frequently in mathematical physics. When $l = 2$, we have the following five spherical harmonics:

$$Y_{20}(\theta,\varphi) = \sqrt{\frac{5}{16\pi}}(3\cos^2\theta - 1),$$

$$Y_{2,-1}(\theta,\varphi) = \sqrt{\frac{15}{8\pi}}e^{-i\varphi}\sin\theta\cos\theta, \qquad Y_{21}(\theta,\varphi) = -\sqrt{\frac{15}{8\pi}}e^{i\varphi}\sin\theta\cos\theta,$$

$$Y_{2,-2}(\theta,\varphi) = \sqrt{\frac{15}{32\pi}}e^{-2i\varphi}\sin^2\theta, \qquad Y_{22}(\theta,\varphi) = \sqrt{\frac{15}{32\pi}}e^{2i\varphi}\sin^2\theta.$$

Using Equation (1.3), make a three-dimensional parametric plot of the real parts of $Y_{21}(\theta,\varphi)$ and $Y_{22}(\theta,\varphi)$. Also make a three-dimensional parametric plot of the square of the absolute values of $Y_{21}(\theta,\varphi)$ and $Y_{22}(\theta,\varphi)$. Recall that $e^{i\alpha} = \cos\alpha + i\sin\alpha$.

Problem 1.25. Using *Mathematica*, find the real and imaginary parts of the following complex numbers:

(a) $(2-i)(3+2i)$ (b) $(2-3i)(1+i)$ (c) $(a-ib)(2a+2ib)$

(d) $\dfrac{i}{1+i}$ (e) $\dfrac{1+i}{2-i}$ (f) $\dfrac{1+3i}{1-2i}$

(g) $\dfrac{1+2i}{2-3i}$ (h) $\dfrac{2}{1-3i}$ (i) $\dfrac{1-i}{1+i}$

(j) $\dfrac{5}{(1-i)(2-i)(3-i)}$ (k) $\dfrac{1+2i}{3-4i} + \dfrac{2-i}{5i}$

Problem 1.26. Using *Mathematica*, convert the following complex numbers to polar form (i.e., find the absolute values and the arguments of the numbers):

(a) $2-i$ (b) $2-3i$ (c) $3-2i$ (d) i

(e) $-i$ (f) $\dfrac{i}{1+i}$ (g) $\dfrac{1+i}{2-i}$ (h) $\dfrac{1+3i}{1-2i}$

(i) $1+i\sqrt{3}$ (j) $\dfrac{2+3i}{3-4i}$ (k) $27i$ (l) -64

(m) $2-5i$ (n) $1+i$ (o) $1-i$ (p) $5+2i$

Problem 1.27. Using *Mathematica*, find the real and imaginary parts of the following:

(a) $(1+i\sqrt{3})^3$ (b) $(2+i)^{53}$ (c) $\sqrt[4]{i}$ (d) $\sqrt[3]{1+i\sqrt{3}}$

(e) $(1+i\sqrt{3})^{63}$ (f) $\left(\dfrac{1-i}{1+i}\right)^{81}$ (g) $\sqrt[6]{-i}$ (h) $\sqrt[4]{-1}$

(i) $\left(\dfrac{1+i\sqrt{3}}{\sqrt{3}+i}\right)^{217}$ (j) $(1+i)^{22}$ (k) $\sqrt[6]{1-i}$ (l) $(1-i)^4$.

Problem 1.28. In Equation (1.5), let $N\phi = \beta$, and use small-angle approximation for the sine term in the denominator. Absorb any constant (such as N^2) in the constant called I_0. To find the expression for β in terms of θ, simply note that $Nd = a$.

2

Vectors and Matrices in *Mathematica*

Now that we have become familiar with the essentials of *Mathematica*, let us apply our knowledge to solving some physical problems. We shall concentrate on specific examples and introduce new techniques as we progress. Most of the examples are based on the author's book, *Mathematical Methods* [Hass 00], which we abbreviate as *MM*, and refer the reader—in the margins—to its appropriate sections and pages for further details and deeper understanding. Although a familiarity with the concepts in that book is helpful for a fuller understanding of our examples, no prior knowledge of those concepts is essential for the *Mathematica* applications.

2.1 Electric Fields

The electric field $\mathbf{E}$ of a point charge q_1 with position vector $\mathbf{r}_1$ at a point P—called the *field point*—with position vector $\mathbf{r}$ is given by the formula

MM, p. 24

$$\mathbf{E} = \frac{k_e q_1}{|\mathbf{r} - \mathbf{r}_1|^3}(\mathbf{r} - \mathbf{r}_1) \tag{2.1}$$

As an example of a vector manipulation, let's have *Mathematica* calculate the field of a single charge for us.

First, we define our position vectors:

how to type in vectors

```
In[1]:= r={x,y,z}; r1={x1,y1,z1};
```

Note that because of the semicolons, *Mathematica* does not give any output. Vectors are always defined as a list, separated by commas and enclosed in curly brackets. One can define vectors of any dimension. All one needs to do is type in more components. As a check that *Mathematica* remembers $\mathbf{r}$ and $\mathbf{r}_1$, we can ask it to calculate $\mathbf{r} - \mathbf{r}_1$:

```
In[2]:= r-r1
Out[2]:= {x-x1,y-y1,z-z1}
```

which are the components of the the desired vector. *Mathematica* can also calculate the dot product of the two vectors

dot product of two vectors

```
In[3]:= r.r1
Out[3]:= xx1+yy1+zz1
```

or

```
In[4]:= (r-r1).(r-r1)
```

Out[4]:= $(x - x1)^2 + (y - y1)^2 + (z - z1)^2$

Some common operations with vectors are collected below

`{a,b,c}`	the vector $\langle a, b, c\rangle$
`a v`	multiply the vector $\mathbf{v}$ by the scalar a
`v.w`	the dot product $\mathbf{v} \cdot \mathbf{w}$
`Cross[v,w]`	the cross product $\mathbf{v} \times \mathbf{w}$
`Array[v,n]`	build an n-dimensional vector of the form $\{v[1], v[2], \ldots, v[n]\}$
`Table[f[k],{k,n}]`	build an n-dimensional vector by evaluating f with $k = 1, 2, \ldots, n$

Next, we write the expression for the field (with $k_e = 1$):

```
In[5]:= EField1[r_,r1_,q1_]:=
           (q1/((r-r1).(r-r1))^(3/2)) (r-r1)
```

Note the occurrence of the "blank" (_) on the left-hand side of the assignment relation. Because of the presence of $(\mathbf{r} - \mathbf{r}_1)$ on the right-hand side, `EField1` will be a vector—which is a function of vectors. If we type

```
In[6]:= EField1[r,r1,q1]
```

the output will be a list of three terms whose first entry is

$$\frac{q1(x - x1)}{((x - x1)^2 + (y - y1)^2 + (z - z1)^2)^{3/2}}$$

with similar entries for the y- and z-components.

We can confine the charge and the field point to the xy-plane by setting the third components of $\mathbf{r}$ and $\mathbf{r}_1$ equal to zero:

```
In[7]:= EField1[{x,y,0},{1,1,0},1]
```

where we have also set $x1 = 1 = y1$ and $q1 = 1$. Note how the argument of `EField1` matches its definition in `In[5]`: The first and second entries are *vectors*, and thus have to be lists with three components.

The output of `In[7]` will be

$$\left\{\frac{-1+x}{((-1+x)^2+(-1+y)^2+(-1+z)^2)^{3/2}}, \frac{-1+y}{((-1+x)^2+(-1+y)^2+(-1+z)^2)^{3/2}}, 0\right\}$$

showing that the third component of the field is zero, as expected.

We now take only the first two components of the field and try to make a two-dimensional plot of the field lines. Here is how it is done:

```
In[8]:= {E2D1x, E2D1y}=Take[EField1[{x,y,0},{1,1,0},1],2];
```

where `Take[list,n]` takes the first n elements of *list* and makes a new list from them. The left-hand side is the name we have chosen for the two-dimensional electric field.

choosing components of a vector using **Take**

Now we are ready to plot the two-dimensional field lines. But first we have to load the correct `Graphics` package. For this, you need to type `<<` followed by the name of the package. For example,

loading packages

```
In[9]:= << Graphics`PlotField`
```

or

```
In[9]:= << Graphics/PlotField.m
```

the subpackage **PlotField** and the command **PlotVectorField**

will load the `PlotField` subpackage, which is needed to plot field lines. Once the package is loaded, we plot the electric field lines using the command

```
In[10]:= PlotVectorField[{E2D1x, E2D1y},{x, 0,2},
          {y, 0,2}, Axes->True]
```

By default, `PlotVectorField` does not draw the axes. To see the axes, one uses the option `Axes->True`. The result is displayed in Figure 2.1. Note that E2D1x and E2D1y are—as they should be for `PlotVectorField` command to work—functions of x and y. The arrows have lengths that are proportional to the strength of the field: close to the charge, they are long, and get shorter and shorter as we move farther and farther away from the charge.

Now let us look at a slightly more complicated field—that of two charges. First, type in the location of the second charge:

```
In[11]:= r2={x2,y2,z2};
```

Then calculate the field due to this charge at the same *field* point as before:

```
In[12]:= EField2[r_,r2_,q2_]:=
         (q2/((r-r2).(r-r2))^(3/2)) (r-r2)
```

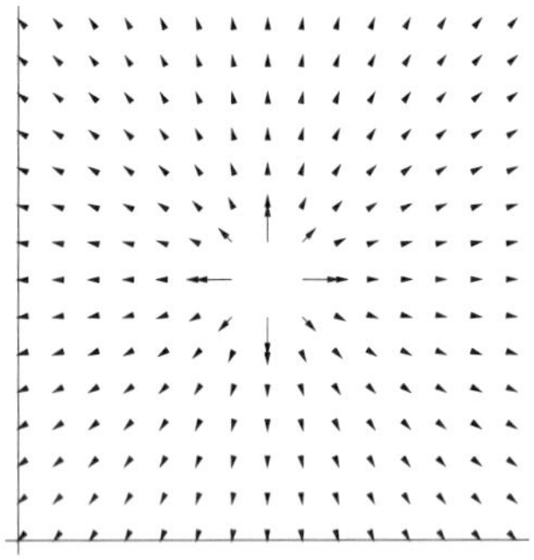

FIGURE 2.1. The field lines of a point charge located at (1, 1).

Now add the two fields to get the total electric field at **r**:

```
In[13]:= Etotal[r_,r1_,r2_,q1_,q2_]:=
         EField1[r,r1,q1]+EField2[r,r2,q2]
```

MM, pp. 25 and 26

Let us see what the field lines of a dipole look like. A dipole is the combination of two charges of equal strength and opposite signs.[1] Let the positive charge of +1 be at $(1, 1, 0)$ and the negative charge of -1 be at $(1, 1.5, 0)$. Let us also assume that the field point is at $(x, y, 0)$. Since we are interested in a two-dimensional plot of the field lines, we separate the first two components of `Etotal`:

```
In[14]:= {EDipx, EDipy}=
         Take[Etotal[{x,y,0},{1,1,0},{1,1.5,0},1,-1],2];
```

Once again the entries of `Etotal` must match those of its definition in `In[13]`. Now we are ready to plot this field. Typing in the input line

```
In[15]:= PlotVectorField[{EDipx, EDipy},{x, 0,2},
         {y, 0,2}, Axes->True, Ticks->None]
```

yields the plot on the left of of Figure 2.2.

We can change the relative strength of the two charges. For example,

```
In[16]:= {E2Qx, E2Qy}=
         Take[Etotal[{x,y,0},{1,1,0},{1,1.5,0},4,-1],2];
```

is the two-dimensional field of two charges, in which the positive charge has four times the strength of the negative charge. Then

```
In[17]:= PlotVectorField[{E2Qx, E2Qy},{x, 0,2},
         {y, 0,2}, Axes->True, Ticks->None]
```

produces the plot on the right of Figure 2.2.

the package **PlotField3D**

Mathematica can also plot the field lines in three dimensions. Simply

[1] Actually, to be precise, we have to define a dipole as two *distant* charges of opposite sign located very close to one another.

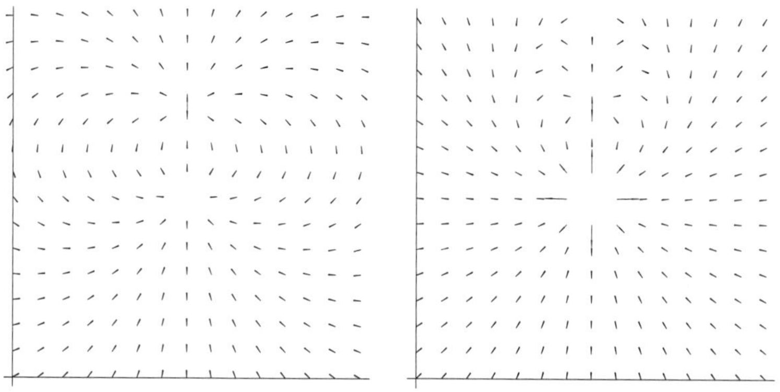

FIGURE 2.2. The field lines of (left) a dipole whose positive charge is located at (1, 1)—the center of the plot—and whose negative charge is located at (1, 1.5), and (right) of two opposite charges, with the positive charge being four times the negative charge.

load the appropriate package using

```
In[18]:= << Graphics`PlotField3D`
```

and type in the three components of the field. For example,

```
In[19]:= {E3Dx,E3Dy,E3Dz}=EField[{x,y,z},{0,0,0},1];
```

where `EField`, defined in `In[5]` on page 50, describes the three components [at (x, y, z)] of a positive charge of unit strength located at the origin. To produce a three-dimensional plot of this field, with the range of x, y, and z confined to the interval $(-1, 1)$, we type in

the command **PlotVectorField3D**

```
In[20]:= PlotVectorField3D[{E3Dx,E3Dy,E3Dz},{x,-1,1},
         {y,-1,1}, {z,-1,1}]
```

and obtain the plot on the left in Figure 2.3. Similarly,

```
In[21]:= {E3Ddipx,E3Ddipy,E3Ddipz}=
         EField[{x,y,z},{0,0,-0.5},-1] +
         EField[{x,y,z},{0,0,0.5},1];
```

corresponding to a dipole on the z-axis, along with

```
In[22]:= PlotVectorField3D[{E3Ddipx,E3Ddipy,E3Ddipz},
         {x, -1,1},{y, -1,1}, {z, -1,1}]
```

will produce the plot on the right in Figure 2.3.

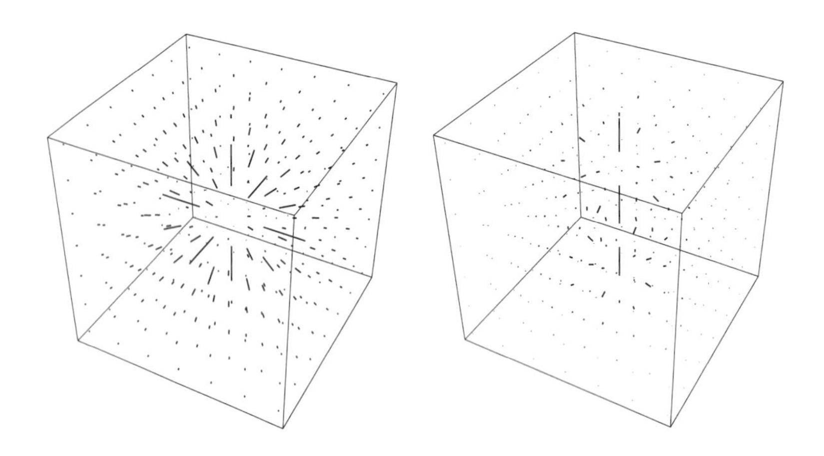

FIGURE 2.3. The field lines of a charge (left) and a dipole (right) in three dimensions.

2.2 Ionic Crystals

The last section discussed vector calculations in *Mathematica* in the setting of discrete charges and their fields. This section considers another aspect of a collection of discrete charges: the energy stored in assembling them together.

Consider a collection of N charges brought together so that the ith particle of charge q_i is located at the position vector $\mathbf{r}_i$. Then, the (potential) energy of the assemblage is given by

MM, pp. 314–315

$$U = \frac{1}{2}\sum_{i=1}^{N} q_i \Phi_i, \qquad \Phi_i = k_e \sum_{j=1}^{N} \frac{q_j}{|\mathbf{r}_i - \mathbf{r}_j|} \tag{2.2}$$

where Φ_i is the electrostatic potential of all the charges (except the ith charge) at the location of the ith charge.

2.2.1 One-Dimensional Crystal

Now suppose that we have $N/2$ positive and an equal number of negative charges arranged alternately on a straight line spaced a distance a apart as shown in Figure 2.4. This arrangement is called a (one-dimensional) **ionic crystal**. Ordinary salt is a three-dimensional collection of positive sodium ions interspersed between the negative chlorine ions.

ionic crystals

We want to calculate the potential energy of the one-dimensional crystal. In a real crystal N is so large that we can consider it to be infinite. Let

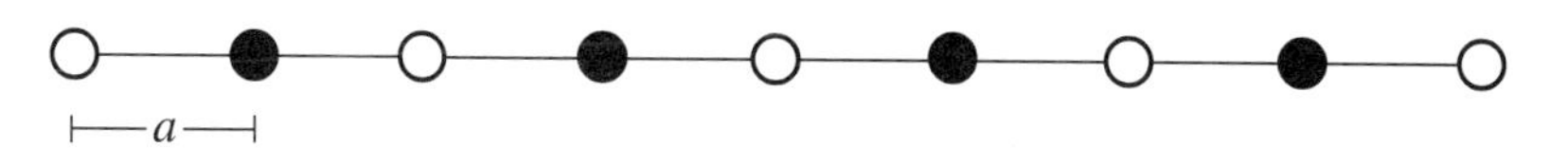

FIGURE 2.4. A one-dimensional ionic crystal.

us suppose that the jth charge has $(-1)^j$ as its sign and ja as its location on the common axis, where j takes on values from $-N/2$ to $+N/2$. This labeling corresponds to taking the "zeroth" charge to be at the origin, with $N/2$ charges on the right and on the left. For this to happen, we actually need $N+1$ charges; but as N is assumed to be very large, $N \approx N+1$.

To find U, we first calculate the electrostatic potential Φ_0 at the origin. Here $\mathbf{r}_0 = 0$ and $\mathbf{r}_j = ja$. Furthermore, the arrangement of charges on the right of the origin is identical to the left. Therefore,

$$\Phi_0 = k_e \sum_{j=-N/2}^{N/2} \frac{q(-1)^j}{|ja|} = 2k_e \sum_{j=1}^{N/2} \frac{q(-1)^j}{ja}$$

and the product of the potential and the charge at the origin is

$$q_0\Phi_0 = \frac{2k_e q^2}{a} \sum_{j=1}^{N/2} \frac{(-1)^j}{j}$$

This shows that if we had chosen a negative charge at the origin, the result would have been the same because the final result depends on q^2. Moreover, because the crystal is assumed infinite in length, all charge locations are identical to the origin. This means that

$$q_0\Phi_0 = q_1\Phi_1 = q_2\Phi_2 = q_3\Phi_3 = \dots$$

i.e., all the terms in the sum (2.2) are equal. Hence,

$$U = \tfrac{1}{2} N q_0 \Phi_0 = N \frac{k_e q^2}{a} \sum_{j=1}^{N/2} \frac{(-1)^j}{j} \quad \text{or} \quad u = \frac{2k_e q^2}{a} \sum_{j=1}^{N/2} \frac{(-1)^j}{j}$$

where u is the potential energy per *molecule*—consisting of two ions, thus the factor of 2. This is usually written as

$$u = -\alpha \frac{k_e q^2}{a}, \quad \text{where} \quad \alpha \equiv -2 \sum_{j=1}^{N/2} \frac{(-1)^j}{j} \tag{2.3}$$

and α is called the **Madelung constant**. Let us employ *Mathematica* to calculate the Madelung constant for a one-dimensional ionic crystal.

Madelung constant in one dimension

This calculation involves simply evaluating the sum in Equation (2.3), for which we type in

```
In[1]:= alph[n_]:= 2 Sum[(-1)^(j+1)/j,{j,1,n}]
```

where we have included the negative sign outside in the exponent of (-1), and used n for $N/2$. For all practical purposes, n is infinite. Thus, we have to take n to be very large. For series that are rapidly convergent, "very large" could be 50, 100, or 500. However, the series above happens to be converging rather slowly, as Table 2.1 shows.

Typing in `N[alph[n]]` for various n, *Mathematica* yields the entries of Table 2.1. The "`N`" in the command is necessary if a decimal representation of the answer is desired. Without it, a rational representation—with large integers in the numerator and denominator—will be outputted. The next-to-last entry is obtained by typing in `N[alph[Infinity]]` and the last entry by typing in `alph[Infinity]`. The summation in (2.3) is a well-known series in calculus, and *Mathematica* recognizes this series.

MM, pp. 225 and 235

2.2.2 Two-Dimensional Crystal

With a little more effort, one can find the Madelung constant for a more realistic two-dimensional crystal. Once again, we place a positive charge at the origin, invoke the symmetry of the crystal, and use $U = \frac{1}{2}Nq\Phi_0$ to calculate the potential energy of N charges arranged symmetrically in two dimensions on squares of side a as shown in Figure 2.5.

It is convenient to designate the charges and their positions with a double index. Thus, q_{ij} is the charge located at a point P_{ij} with coordinates (ia, ja) in the xy-plane, and Φ_{ij} is the electrostatic potential at P_{ij} due to all other charges. The potential energy of a crystal is therefore

$$U = \tfrac{1}{2}\sum_{i,j} q_{ij}\Phi_{ij} = \tfrac{1}{2}Nq\Phi_{00} \tag{2.4}$$

where we have used the symmetry of the crystal, i.e., that $q_{ij}\Phi_{ij}$ is the same for all sites, and therefore, the sum is simply N times one of the terms, which we have chosen to be that corresponding to the origin. Denoting the position vector of P_{jk} by $\mathbf{r}_{jk}$, we have

$$\Phi_{00} = \sum_{j=-n}^{n}\sum_{k=-n}^{n} \frac{k_e q_{jk}}{|\mathbf{r}_{jk}|} = \sum_{j=-n}^{n}\sum_{k=-n}^{n} \frac{k_e q(-1)^{j+k}}{\sqrt{(ja)^2 + (ka)^2}}$$

where we have used $q_{ij} = q(-1)^{j+k}$ corresponding to a positive charge at the origin with other charges alternating in sign in both x and y directions. Here n is a large number related to N. In fact, the reader may verify that $n = \sqrt{N}/2$. Substituting the result above in Equation (2.4) gives

$$U = \tfrac{1}{2}N\frac{k_e q^2}{a}\sum_{j=-n}^{n}\sum_{k=-n}^{n} \frac{(-1)^{j+k}}{\sqrt{j^2 + k^2}} \tag{2.5}$$

n	alph(n)	n	alph(n)
20	1.33754	5000	1.38609
50	1.36649	10000	1.38619
100	1.37634	15000	1.38623
500	1.38430	20000	1.38624
1000	1.38529	30000	1.38626
2000	1.38579	∞	1.38629
4000	1.38604	∞	2 Log[2]

TABLE 2.1. The values of the sum in Equation (2.3) for increasing number of terms of the sum.

Let us first sum over k. This can be split into three pieces:

$$\sum_{k=-n}^{n} = \sum_{k=-n}^{-1} + (k = 0 \text{ term}) + \sum_{k=1}^{n}$$

The first and last pieces give identical results, because the summand is a function of k^2, which is insensitive to the sign of k. We thus have

$$\begin{aligned} U &= \tfrac{1}{2}N\frac{k_e q^2}{a}\sum_{j=-n}^{n}\left\{\frac{(-1)^j}{\sqrt{j^2}} + 2\sum_{k=1}^{n}\frac{(-1)^{j+k}}{\sqrt{j^2+k^2}}\right\} \\ &= \tfrac{1}{2}N\frac{k_e q^2}{a}\left\{\sum_{j=-n}^{n}\frac{(-1)^j}{\sqrt{j^2}} + 2\sum_{j=-n}^{n}\sum_{k=1}^{n}\frac{(-1)^{j+k}}{\sqrt{j^2+k^2}}\right\} \\ &= \tfrac{1}{2}N\frac{k_e q^2}{a}\left\{2\sum_{j=1}^{n}\frac{(-1)^j}{j} + 4\sum_{j=1}^{n}\sum_{k=1}^{n}\frac{(-1)^{j+k}}{\sqrt{j^2+k^2}} + 2\sum_{k=1}^{n}\frac{(-1)^k}{\sqrt{k^2}}\right\} \end{aligned}$$

where we split the sum over j as we did for k. The first and third terms in the curly brackets of the last line are equal. It follows that

Madelung constant in two dimensions

$$u = -\alpha\frac{k_e q^2}{a} \quad \text{where} \quad \alpha \equiv -4\sum_{j=1}^{n}\frac{(-1)^j}{j} - 4\sum_{j=1}^{n}\sum_{k=1}^{n}\frac{(-1)^{j+k}}{\sqrt{j^2+k^2}} \tag{2.6}$$

where u is, as before, the potential energy (or *binding* energy, as it is usually called) per ion, of which there are $N/2$.

To find a numerical estimate for the two-dimensional Madelung constant, type in

```
In[1]:= Mad2D[n_]:= -4 Sum[(-1)^j/j,{j,1,n}]
        -4 Sum[(-1)^(j+k)/Sqrt[j^2+k^2],{j,1,n},{k,1,n}]
```

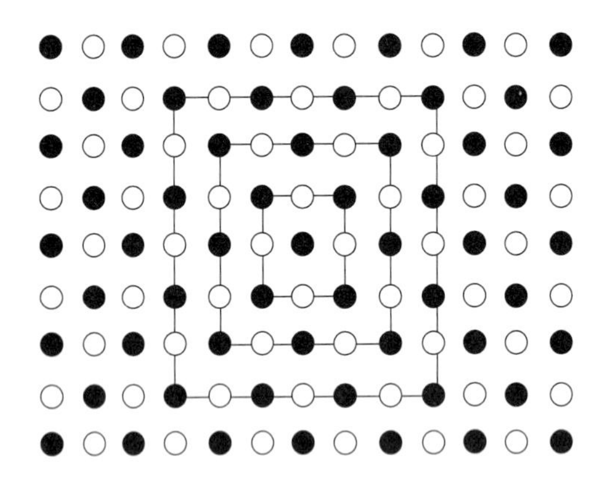

FIGURE 2.5. A two-dimensional ionic crystal with equidistant ions in both perpendicular directions. One can also calculate the potential energy by adding contributions of squares (three of which are shown) centered around the origin. We shall not pursue such a calculation.

where n is a very large number. Note that *Mathematica* can do double (or nested) sums. Because of the double sum, it will take *Mathematica* longer to evaluate `Mad2D[n]` for large values of n, as the number of calculations involved is n^2. Table 2.2 shows some sample results.

2.2.3 Three-Dimensional Crystal

The real, three-dimensional Madelung constant can now be calculated using our experience with the two-dimensional case. We can actually start as we did in the case of two dimensions and write the potential as a triple sum, the generalization of Equation (2.5). Then split the sum into three pieces corresponding to negative values, zero, and positive values of the summation index. However, we can use the two-dimensional result as a guide to facilitate the present calculation.

geometric interpretation of 2D Madelung constant

The sums in the Madelung constant α in Equation (2.6) have a plausible geometric explanation. The first sum calculates the contribution to the binding energy of the charges on the axes. The factor 4 counts the number of *semi*axes: right, left, up, and down. The second sum calculates the contribution of the charges *off the axes* in the four quadrants.

This geometric interpretation can be easily generalized to three dimensions. There must be six single sums corresponding to the positive and negative sides of each axis. Now each of the three planes has four quadrants. Therefore, we expect 12 double summations. Finally, the remaining charges will occupy the eight octants, leading to eight triple sums. It now follows that for the three-dimensional ionic crystal, we should have $u = -\alpha\frac{k_e q^2}{a}$ with

3D Madelung constant

n	Mad2D(n)	n	Mad3D(n)
10	1.54824	20	1.7194
20	1.58105	30	1.72864
100	1.60851	40	1.73331
200	1.61202	60	1.73802
400	1.61378		

TABLE 2.2. Some approximations for the Madelung constant in two and three dimensions.

$$\alpha \equiv -6\sum_{j=1}^{n}\frac{(-1)^j}{j} - 12\sum_{j=1}^{n}\sum_{k=1}^{n}\frac{(-1)^{j+k}}{\sqrt{j^2+k^2}} - 8\sum_{i=1}^{n}\sum_{j=1}^{n}\sum_{k=1}^{n}\frac{(-1)^{i+j+k}}{\sqrt{i^2+j^2+k^2}} \tag{2.7}$$

where n is again a very large number (equal to $\sqrt[3]{N}/2$).

To find a numerical estimate for the three-dimensional Madelung constant, type in

```
In[1]:= Mad3D[n_]:= -6 Sum[(-1)^j/j,{j,1,n}]
          -12 Sum[(-1)^(j+k)/Sqrt[j^2+k^2],{j,1,n},{k,1,n}]
          -8 Sum[(-1)^(i+j+k)/Sqrt[i^2+j^2+k^2],{i,1,n},
            {j,1,n},{k,1,n}]
```

Because of the triple sum, it will take *Mathematica* much longer to evaluate `Mad3D[n]` for large values of n. In fact, the time is proportional to n^3. Thus, for a modest $n = 100$, the time required is 10,000 times the duration of a one-dimensional calculation with the same number of ions. Table 2.2 shows some sample results.

Remarks: A few comments on the *physics* of the ionic crystals are in order. First we note that there is little difference among the Madelung constants in one, two, or three dimensions. Since this determines the bonding of the molecules, we conclude that the binding energy per molecule is not terribly sensitive to the dimensionality of the crystal. Next, in all cases, the Madelung constant is positive, leading to a *negative* energy per ion. This indicates that the binding energy is negative. What is the significance of this? Imagine taking the ions and separating them so far that they would not feel the electrostatic forces of one another. Such a state would correspond to a zero potential energy. If the ions are moving at all, then their total energy, which is equal to the total kinetic energy, will be positive, and the energy per ion (or molecule) will be positive.

Conservation of energy now tells us that, if we wish to "break up" the crystal completely—i.e., send the ions infinitely far from one another—we have to supply each ion with some positive energy to overcome its negative binding energy. With $k_e = 9 \times 10^9$, $q = 1.6 \times 10^{-19}$, and a a typical molecular distance (a few Ångstrøm), the binding energy per ion will be about 5×10^{-19} Joule. With a typical crystal sample having approximately 10^{30} molecules, we see that a tremendous amount of energy, something like 5×10^{11} Joules, is required to *completely* dissociate the crystal. On the other hand, if we are interested in simply breaking the crystal *in half*, we need to overcome the binding energy of the ions at the interface of the two resulting halves. This amounts to the binding energy of about 10^{20} molecules or 50 Joules, something quite manageable.

The Madelung constant has been the subject of intense numerical study, and many series representations of this constant are by far superior to the series discussed in this section [Cran 96, pp. 73–79]. One such representation for the Madelung constant in three dimensions is

$$\alpha = 12\pi \sum_{j=0}^{\infty} \sum_{k=0}^{\infty} \operatorname{sech}^2\left(\frac{\pi}{2}\sqrt{(2j+1)^2 + (2k+1)^2}\right) \tag{2.8}$$
$$= 1.7475645946331821906362120355443974034851614366247\ldots$$

Even using 2 instead of infinity in the double sum above will give the three-dimensional Madelung constant to six significant figures.

2.3 Tubing Curves

This section deals with a topic that from the standpoint of physics is irrelevant, but from a *Mathematica* standpoint is quite interesting. The problem treated here is to find a "tube" that follows the shape of a curve with a given parametric equation. Specifically, suppose that the parametric equation of a curve is given by

$$x = f(t), \quad y = g(t), \quad z = h(t)$$

We are interested in finding the equation of the *surface* with a circular cross section of radius a that surrounds this curve [Figure 2.6(a)].

A point P on the desired surface can be reached by adding two "natural" vectors: a vector $\mathbf{r}(t)$ that connects the origin to a point on the curve, and a vector $\mathbf{a}(u)$ that connects that point to a point on the surface. If $\mathbf{R}$ denotes the position vector of P, then

$$\mathbf{R}(t,u) \equiv \langle X(t,u), Y(t,u), Z(t,u)\rangle = \mathbf{r}(t) + \mathbf{a}(u)$$

The first vector $\mathbf{r}(t)$ has components $\langle f(t), g(t), h(t)\rangle$, corresponding to the parameterization of the curve, and the second vector has components

$$\langle a\cos u \cos\varphi, a\cos u \sin\varphi, a\sin u\rangle \tag{2.9}$$

where u is the angle that $\mathbf{a}(u)$ makes with the cylindrical unit vector $\hat{\mathbf{e}}_\rho$, and φ is the angle that $\hat{\mathbf{e}}_\rho$ makes with the positive x-axis [see Figure 2.6(b)].

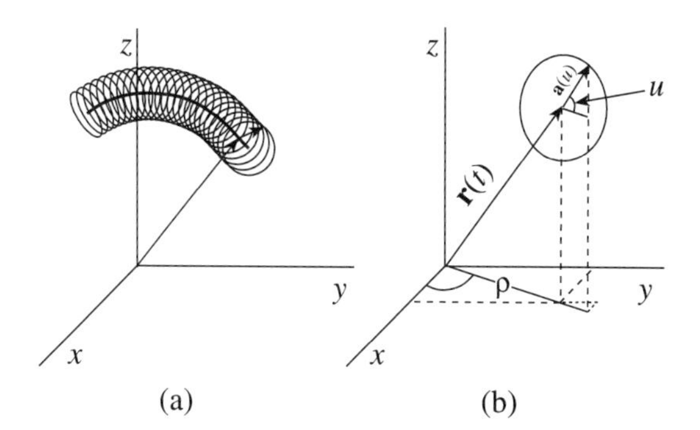

FIGURE 2.6. (a) A curve with its surrounding tube. (b) Details of circular cross section.

We need to express the coordinates of P as a function of two parameters, which we choose to be t and u. To this end we find $\cos\varphi$ and $\sin\varphi$ in terms of t:

$$\cos\varphi = \frac{x}{\rho} = \frac{f(t)}{\sqrt{f^2(t)+g^2(t)}}, \quad \sin\varphi = \frac{y}{\rho} = \frac{g(t)}{\sqrt{f^2(t)+g^2(t)}}$$

It now follows that

$$\begin{aligned} X(t,u) &= f(t) + a\cos u\cos\varphi = f(t)\left(1+\frac{a\cos u}{\sqrt{f^2(t)+g^2(t)}}\right) \\ Y(t,u) &= g(t) + a\cos u\sin\varphi = g(t)\left(1+\frac{a\cos u}{\sqrt{f^2(t)+g^2(t)}}\right) \\ Z(t,u) &= h(t) + a\sin u \end{aligned} \tag{2.10}$$

To create some plots, we type in these equations in *Mathematica*:

```
In[1]:= X[t_,u_,a_]:= f[t](1+aCos[u]/Sqrt[f[t]^2+g[t]^2])

In[2]:= Y[t_,u_,a_]:= g[t](1+aCos[u]/Sqrt[f[t]^2+g[t]^2])

In[3]:= Z[t_,u_,a_]:= h[t] + a Sin[u]
```

First let's make a torus, for which the curve is a circle in the xy-plane.

```
In[4]:= f[t_]:= 3 Cos[t]; g[t_]:= 3 Sin[t]; h[t_]:= 0;
```

We also take the radius of the cross section to be 1 (i.e., $a = 1$). To create the plot, we use

Axes and **Boxed** options for 3D plots

```
In[5]:= ParametricPlot3D[{X[t,u,1],Y[t,u,1],Z[t,u,1]},
         {t,0,2 Pi},{u,0,2 Pi}, Axes->False, Boxed->False]
```

FIGURE 2.7. A torus, a clover leaf tube, and a helix.

`Axes->False` and `Boxed->False` will eliminate the axes and the box surrounding the plot, respectively. The result is the graph displayed on the left in Figure 2.7.

Next we produce a tubing of a clover leaf. We need to clear all functions that are to be inserted in the components of `ParametricPlot3D`. This is necessary, because *Mathematica* will not "forget" the definitions of functions unless you explicitly remove them using `Clear`:

```
In[6]:= Clear[f,g]
```

We still want h to be zero. Now define the f and g corresponding to a clover leaf:

```
In[7]:= f[t_]:=4 Cos[t] Cos[2t]; g[t_]:=4 Sin[t] Cos[2t];
```

Instead of 4 multiplying the trigonometric functions, we could have chosen any other constant. This constant determines the length of each leaf. To create the plot, we once again use

```
In[8]:= ParametricPlot3D[{X[t,u,1],Y[t,u,1],Z[t,u,1]},
        {t,0,2 Pi},{u,0,2 Pi}, PlotPoints->25,Axes->False,
        Boxed->False]
```

PlotPoints option for 3D plots

We used `PlotPoints->25` to make the plot a little smoother. The default value of 15 was fine for the torus, because a torus is intrinsically smooth. The result is displayed in the center of Figure 2.7.

The last plot we want to produce is a helical tube. We first clear the functions:

```
In[9]:= Clear[f,g,h]
```

This time we are clearing h because it is nonzero for a helix. Now we type in the functions describing a helix,

```
In[10]:= f[t_]:= 3 Cos[t]; g[t_]:= 3 Sin[t];
         h[t_]:= 0.5 t;
```

and increase the plot points to 40 for smoother graphics:

```
In[11]:= ParametricPlot3D[{X[t,u,1],Y[t,u,1],Z[t,u,1]},
         {t,0,6 Pi},{u,0,2 Pi}, PlotPoints->40,Axes->False,
         Boxed->False]
```

The result is displayed on the right in Figure 2.7. By replacing the functions f, g, and h with any other triplet, one can produce a variety of tubes.

Let us examine more closely what we have done so far. It is clear from our discussion that what determines the surface is the vector $\mathbf{a}(u)$. Equation (2.9) shows that the vector sum of the first two components of $\mathbf{a}(u)$ is a vector in the $\hat{\mathbf{e}}_\rho$-direction. Therefore, $\mathbf{a}(u)$ is in the plane formed by $\mathbf{r}(t)$ and the z-axis. So, the circular cross sections are not really circular. A true circular cross section should describe a circle *perpendicular* to the curve. Because $\mathbf{r}(t)$ and the z-axis are automatically perpendicular to the circular curve generating the torus, the cross section is automatically perpendicular to that curve. However, the other two curves do not satisfy this property, although this fact is less clear for the helix. But the clover leaf shows this lack of circularity in a pronounced way, especially close to the origin.

To make the circular cross section perpendicular to the curve, we have to choose $\mathbf{a}(u)$ to be of constant length and perpendicular to the curve at its point of contact.[2] This means that $\mathbf{a}(u)$ is to be perpendicular to an element of length $d\mathbf{r}$ of the curve. But

$$d\mathbf{r} = \langle df, dg, dh \rangle = \langle f'(t), g'(t), h'(t) \rangle \, dt$$

So $\mathbf{a}(u)$ must be perpendicular to $\mathbf{r}'(t) \equiv \langle f'(t), g'(t), h'(t) \rangle$. Letting $\mathbf{a}(u)$ have components s, v, and w, we have our first relation:

$$\mathbf{r}'(t) \cdot \mathbf{a}(u) = 0 = \langle f'(t), g'(t), h'(t) \rangle \cdot \langle s, v, w \rangle$$

We therefore start our *Mathematica* program by typing in

```
In[1]:= r[t_]:={f[t],g[t],h[t]}; rp={s,v,w};

In[2]:= r'[t].rp==0
```

where we have used `rp` for $\mathbf{a}(u)$.[3] Recall that equations (rather than *assignments*) require two equal signs in *Mathematica*, and that *Mathematica* is familiar with prime (′) representing the derivative of a function of a single variable.

As before [see Figure 2.6(b)], we want to define an angle u on the circular cross section to measure the direction of $\mathbf{a}(u)$. For this, we need a fiducial

---

[2]The variable u is as yet not defined but will be chosen below.

[3]We cannot use `a[u]`, because *Mathematica* would expect `a` to have been defined as a function of u. We do not want to use a, because we want to reserve a for the radius of the cross section.

axis relative to which we measure the angle. This axis must lie on the plane of the circle, of course [and, therefore, must be perpendicular to the curve, i.e., to $\mathbf{r}'(t)$], but is otherwise arbitrary. A vector that defines such an axis could be taken to be $\langle h'(t), 0, -f'(t)\rangle$, as the reader may easily verify. Thus, we get our second equation:

```
In[3]:= horV:={h'[t],0,-f'[t]};

In[4]:= Cos[u]==rp.horV/(a Sqrt[h'[t]^2+f'[t]]^2);
```

where a is the length of the vector `rp`, which gives us the last equation we need:

```
In[5]:= rp.rp==a^2;
```

`In[4]` uses the definition of the ordinary dot product to find the (cosine of the) angle between $\mathbf{a}(u)$ and the fiducial axis.

The three equations in `In[2]`, `In[4]`, and `In[5]` could be solved to yield s, v, and w in terms of u. Thus, we ask *Mathematica* to do so:

```
In[6]:= Solve[{%2,%4,%5},{s,v,w}]
```

The result is a list of solutions with very long expressions. We now simplify the result

```
In[7]:= FullSimplify[%]
```

and get something a little shorter than before. The solution contains two values for each component of `rp`. We want to choose the first one for each. Let us first pick s. This is done by typing in

```
In[8]:= s /. %7
```

This is of the form `expr /. sol` as discussed in Section 1.5. Here `expr` is just `s` and solution has no name but is given (in its simple form) in `Out[7]` (not printed here). The two solutions for s will be put out as a list. Since we want the first choice, we type in

how to get part of a list

```
In[9]:= Part[%,1]
```

where `Part[list,n]` yields the nth element of `list`.[4] The output, after applying `FullSimplify`, will be

$$ah'\left(\frac{\cos u}{\sqrt{f'^2+h'^2}}+\frac{a\sin^2 u f' g'^2}{\sqrt{a^2\sin^2 u g'^2 h'^2 (f'^2+h'^2)(f'^2+g'^2+h'^2)}}\right)$$

To extract the powers out of the radical, we apply `PowerExpand` to the expression. In fact, the input

use of **PowerExpand**

[4]`First[list]` has the same effect as `Part[list,1]`.

```
In[10]:= S[t_,u_,a_]:=PowerExpand[%]
```

defines the function $S(t,u,a)$ as given below:

$$S(t,u,a) = \frac{ah' \cos u}{\sqrt{f'^2 + h'^2}} + \frac{af'g' \sin u}{\sqrt{(f'^2 + h'^2)(f'^2 + g'^2 + h'^2)}} \tag{2.11}$$

This is what we had called s before, but now we have explicitly articulated its dependence on t and u (and a for later use).

To select v, go through the same steps:

```
In[11]:= v /. %7;
```

and

```
In[12]:= Part[%,1]
```

and

```
In[13]:= V[t_,u_,a_]:= PowerExpand[%]
```

to obtain

$$V(t,u,a) = -\frac{a \sin u \sqrt{f'^2 + h'^2}}{\sqrt{f'^2 + g'^2 + h'^2}} \tag{2.12}$$

Finally, the three inputs

```
In[14]:= w /. %7;
```

```
In[15]:= Part[%,1];
```

and

```
In[16]:= W[t_,u_,a_]:= PowerExpand[%]
```

yield

$$W(t,u,a) = -\frac{af' \cos u}{\sqrt{f'^2 + h'^2}} + \frac{ag'h' \sin u}{\sqrt{(f'^2 + h'^2)(f'^2 + g'^2 + h'^2)}} \tag{2.13}$$

What is left now is adding these functions to the components of $\mathbf{r}(t)$ to get the functions needed for the three-dimensional parametric plot:

```
In[17]:= X[t_,u_,a_]:=f[t]+S[t,u,a];
          Y[t_,u_,a_]:=g[t]+V[t,u,a];
          Z[t_,u_,a_]:=h[t]+W[t,u,a];
```

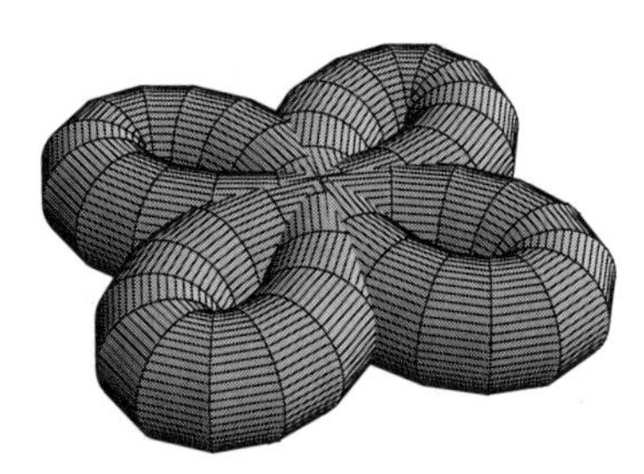

FIGURE 2.8. A tubing of a clover leaf whose cross section is circular.

For plane curves—lying in the xy-plane—we have $h(t) = 0$, and the expressions for S, V, and W simplify considerably. In fact, we have

$$\begin{aligned} X(t,u,a) &= f(t) + \frac{ag' \sin u}{\sqrt{f'^2 + g'^2}} \\ Y(t,u,a) &= g(t) - \frac{af' \sin u}{\sqrt{f'^2 + g'^2}} \\ Z(t,u,a) &= a \cos u \end{aligned} \tag{2.14}$$

Using the new expressions will not affect the torus and will have a small and unnoticeable effect on the helix. But the effect is conspicuous for the clover leaf, so with

```
In[19]:= Clear[f,g,h]
```

and

```
In[20]:= f[t_]:= 4 Cos[t] Cos[2t]; g[t_]:=
           4 Sin[t] Cos[2t]; h[t_]:=0
```

we obtain Figure 2.8, which depicts a circular cross section unlike the clover leaf of Figure 2.7.

2.4 Matrices

MM, Chapter 4

Ever since the discovery of quantum mechanics, matrices have been playing an increasingly important role in physics. It is therefore worth our effort to gain some familiarity with how *Mathematica* handles matrix manipulations.

A matrix is typed in as a list of lists, although the `BasicInput` palette allows typing in the matrix as elements in rows and columns, as is normally done. For example,

```
In[1]:= m= {{a11,a12,a13},{a21,a22,a23},{a31,a32,a33}};
```

is the matrix

$$\begin{pmatrix} a11 & a12 & a13 \\ a21 & a22 & a23 \\ a31 & a32 & a33 \end{pmatrix}$$

and we can verify that by typing in `MatrixForm[m]` for which the output will display the matrix in rows and columns. Of course, one can construct matrices whose numbers of rows and columns are not equal.

use of **MatrixForm**

One can multiply two matrices as long as the matrix on the left has the same number of columns as the one on the right has rows. Thus, with

```
In[2]:= A= {{1,2,3},{-4,-5,6},{7,8,9}};
```

and

```
In[3]:= B= {{-1,1},{1,0},{1,1}};
```

we can type

```
In[4]:= A.B
```

and get

```
Out[4]:= {{4,4},{5,2},{10,16}}
```

but `B.A` will generate an error message conveying the incompatibility of the matrices for the operation of product.

When a matrix has the same number of rows and columns, one can define its *determinant*. Calculating the determinant of a (square) matrix is easy in *Mathematica*. For example, if we are interested in the determinant of the matrix `A` above, we type

MM, pp. 171–176

```
In[5]:= Det[A]
```

and obtain 72 as the result. The determinant is an important property of a matrix. It *determine*s whether the matrix has an inverse. Since multiplication is defined for any two square matrices of equal number of rows and columns, one can ask if a matrix exists whose product with a given matrix yields the identity matrix.[5] It turns out that only if its determinant is nonzero does a matrix have an inverse. We have seen that the determinant of `A` is nonzero. So, let us evaluate its inverse: type in

MM, p. 189

```
In[6]:= invA=Inverse[A]
```

[5]The reader recalls that the identity matrix has 1's on its main diagonal and zero everywhere else.

to get the following output:

$$\left\{\left\{-\frac{31}{24}, \frac{1}{12}, \frac{3}{8}\right\}, \left\{\frac{13}{12}, -\frac{1}{6}, -\frac{1}{4}\right\}, \left\{\frac{1}{24}, \frac{1}{12}, \frac{1}{24}\right\}\right\}$$

To check *Mathematica*'s result, type in

```
In[7]:= invA.A
```

and obtain

```
Out[7]:= {{1,0,0},{0,1,0},{0,0,1}}
```

Similarly, switching the order

```
In[8]:= MatrixForm[A.invA]
```

yields

```
Out[8]:=
```

$$\begin{pmatrix} 1 & 0 & 0 \\ 0 & 1 & 0 \\ 0 & 0 & 1 \end{pmatrix}$$

Some common operations on matrices are collected below:

`{{a,b,c},{d,e,f}}`	the matrix $\begin{pmatrix} a & b & c \\ d & e & f \end{pmatrix}$
`a A`	multiply the matrix **A** by the scalar a
`A.B`	the matrix product of **A** and **B**
`Inverse[A]`	the inverse of the matrix **A**
`Det[A]`	the determinant of the matrix **A**
`Transpose[A]`	the transpose of the matrix **A**
`Part[A,i,j]`	give the ijth element of the matrix **A**

One of the applications of matrices is in solving systems of linear equations. We consider the special case in which the number of equations is equal to the number of unknowns. To be concrete, let's consider the following system of four equations in four unknowns:

$$\begin{aligned} x + y - z + 2w &= 1 \\ 2x - y + z - w &= -1 \\ x + 2y - z + w &= 2 \\ x + y - 2w &= -2 \end{aligned}$$

As the reader may know, this equation can be written in matrix form:

$$\begin{pmatrix} 1 & 1 & -1 & 2 \\ 2 & -1 & 1 & -1 \\ 1 & 2 & -1 & 1 \\ 1 & 1 & 0 & -2 \end{pmatrix} \begin{pmatrix} x \\ y \\ z \\ w \end{pmatrix} = \begin{pmatrix} 1 \\ -1 \\ 2 \\ -2 \end{pmatrix} \quad \text{or} \quad \mathbf{AX} = \mathbf{B}$$

Generally, if $\mathbf{A}$ has an inverse, we can multiply both sides of the *matrix* equation by $\mathbf{A}^{-1}$ to obtain

$$\mathbf{A}^{-1}\mathbf{AX} = \mathbf{A}^{-1}\mathbf{B} \quad \text{or} \quad \mathbf{1X} = \mathbf{A}^{-1}\mathbf{B} \quad \text{or} \quad \mathbf{X} = \mathbf{A}^{-1}\mathbf{B}$$

In the case at hand, all we need to do is to have *Mathematica* calculate $\mathbf{A}^{-1}$. To this end, we enter $\mathbf{A}$ and $\mathbf{B}$:

```
In[1]:= A={{1,1,-1,2},{2,-1,1,-1},{1,2,-1,1},{1,1,0,-2}};
  B={1,-1,2,-2};
```

and quickly obtain the solution:

```
In[2]:= {x,y,z,w}=Inverse[A].B;
```

Typing x yields $-\frac{3}{4}$; and typing y, z, and w yields $\frac{13}{4}$, 6, and $\frac{9}{4}$, respectively.

The procedure above works as long as the matrix of coefficients $\mathbf{A}$ is invertible and the system of equations is inhomogeneous (not all numbers on the right-hand side are zero). If $\mathbf{A}$ is not invertible, the inhomogeneous system may or may not have a solution, but the homogeneous system of equations (where all numbers on the right-hand side are zero) has at least one solution. When $\mathbf{A}$ is invertible, the only solution to the homogeneous system is the trivial (zero) solution. We shall not pursue the theory of systems of linear equations any further.

MM, Section 4.7

For symbolic calculations, *Mathematica* can generate a different variety of matrices with specific elements that can be manipulated. The following is a partial list of such matrices:

`Array[a,{m,n}]`	build an $m \times n$ matrix with the ijth element being $a[i,j]$
`Table[f[i,j],{i,m},{j,n}]`	build an $m \times n$ matrix by evaluating f with i ranging from 1 to m and j ranging from 1 to n
`IdentityMatrix[n]`	generate an $n \times n$ identity matrix
`DiagonalMatrix[list]`	generate a diagonal matrix with the elements in `list` on the diagonal

Generate two 3×3 matrices with elements `a[i,j]` and `b[i,j]`:

```
In[1]:= A=Array[a,{3,3}]; B=Array[b,{3,3}];
```

using **Part** to find an element of a matrix

The element `[2,3]` of the product **AB** can be extracted as follows:

```
In[2]:= Part[A.B,2,3]
```

The result is

```
Out[2]:= a[2,1] b[1,3]+a[2,2] b[2,3]+a[2,3] b[3,3]
```

2.5 Normal Modes

One of the nicest numerical application of matrices is in the calculation of normal modes in solids. A typical solid is a collection of regularly organized (crystal) atoms or molecules interacting electromagnetically among one another. This interaction can be approximated by a harmonic oscillator potential, i.e., one can assume that the constituents of a solid are attached to their neighbors by a spring. Most of the interesting properties of solids are then connected to the various "modes" in which these springs oscillate.

These modes, called **normal modes**, are described by special angular frequencies with which the entire collection of constituents oscillate. These frequencies are obtained by solving a matrix equation that guarantees the existence of solutions to the equations of motion of the constituents of the solid. In order to make the discussion simple, we consider only one-dimensional solids. Thus, we solve the problem of a collection of identical masses attached by identical massless springs along a single straight line, on which the masses are constrained to move.

2.5.1 A System of Two Masses

To begin, consider the simple case of two identical masses m connected by a single spring of spring constant k, as shown in Figure 2.9. Let the position of the first and second mass from some origin be x_1 and x_2, respectively. The equations of motion for the two masses are

$$
\begin{aligned}
m\ddot{x}_1 &= k(x_2 - x_1 - L) \\
m\ddot{x}_2 &= -k(x_2 - x_1 - L)
\end{aligned}
$$

where L is the unstretched length of the spring. The negative sign in the second equation arises because when the spring is stretched (so that $x_2 - x_1 > L$) the direction of the force on the second mass is opposite to the direction of x_2. We want to write the equations above in terms of

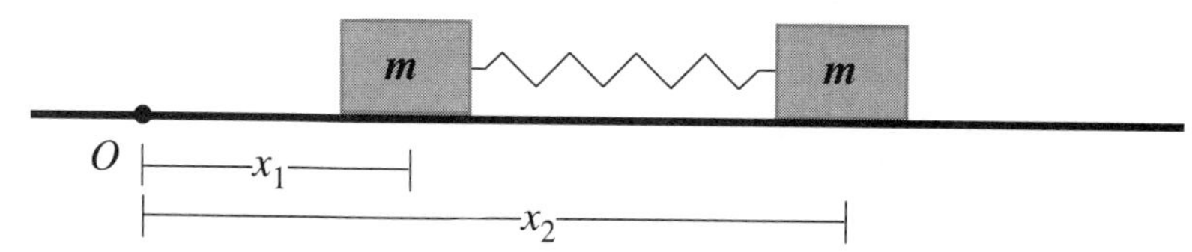

FIGURE 2.9. Two masses attached by a string moving along a straight line on a frictionless horizontal surface.

displacements from equilibrium. Let us call the equilibrium positions of the two masses x_{10} and x_{20}. Then it is clear that $L = x_{20} - x_{10}$; and if we let

$$u_1 \equiv x_1 - x_{10}, \qquad u_2 \equiv x_2 - x_{20}$$

then the equations of motion become

$$\begin{aligned} m\ddot{u}_1 &= k(u_2 - u_1) \\ m\ddot{u}_2 &= -k(u_2 - u_1) \end{aligned} \tag{2.15}$$

because $\ddot{x}_1 = \ddot{u}_1$ and $\ddot{x}_2 = \ddot{u}_2$.

Now we want to solve these equations. From our experience with a single mass, we know that the two masses will execute simple harmonic motions. Thus, both displacements will have the general form

$$u = a \sin \omega t + b \cos \omega t$$

To simplify this further, we assume that at $t = 0$ both displacements are zero. It follows that

$$u_1 = a_1 \sin \omega_1 t \quad \text{and} \quad u_2 = a_2 \sin \omega_2 t$$

Substituting these in Equation (2.15) yields

why the two oscillators must have the same frequency

$$\begin{aligned} (-m\omega_1^2 + k)a_1 \sin \omega_1 t - ka_2 \sin \omega_2 t &= 0 \\ -ka_1 \sin \omega_1 t + (-m\omega_2^2 + k)a_2 \sin \omega_2 t &= 0 \end{aligned} \tag{2.16}$$

These two equations must hold for any value of t. Hence, the coefficient of each sine term must vanish. If $\omega_2 \neq \omega_1$, then the first equation gives $a_2 = 0$, and the second equation yields $a_1 = 0$, a trivial solution. Thus, for nontrivial solutions, we need to assume that $\omega_2 = \omega_1 \equiv \omega$. From Equation (2.16) it follows that

$$\begin{aligned} [(-m\omega^2 + k)a_1 - ka_2] \sin \omega t &= 0 \\ [-ka_1 + (-m\omega^2 + k)a_2] \sin \omega t &= 0 \end{aligned}$$

For this equation to hold for arbitrary t, we must have

$$\begin{aligned} (-m\omega^2 + k)a_1 - ka_2 &= 0 \\ -ka_1 + (-m\omega^2 + k)a_2 &= 0 \end{aligned}$$

or, in matrix form,

$$\begin{pmatrix} k - m\omega^2 & -k \\ -k & k - m\omega^2 \end{pmatrix} \begin{pmatrix} a_1 \\ a_2 \end{pmatrix} = \begin{pmatrix} 0 \\ 0 \end{pmatrix} \tag{2.17}$$

In order to obtain a nontrivial solution to Equation (2.17), we must demand that the 2×2 matrix in that equation not be invertible. This requires that its determinant vanish, yielding a polynomial equation in ω. The roots of this polynomials are the **normal frequencies** corresponding to the normal modes of the system of two masses. Thus, we demand that

$$\det \begin{pmatrix} k - m\omega^2 & -k \\ -k & k - m\omega^2 \end{pmatrix} = 0 \quad \text{or} \quad (k - m\omega^2)^2 - k^2 = 0$$

condition for the existence of solutions

or

$$m^2\omega^4 - 2km\omega^2 = 0$$

whose solution set is $\omega = 0$ and $\omega = \sqrt{2k/m}$, ignoring the negative roots. The zero frequency corresponds to the nonoscillatory motion of the center of mass, and it is usually not counted as a normal mode. Thus, *there is only one normal mode for two masses connected by a spring.*

"Mode" designates more than just the frequency. It describes the manner in which the two masses move relative to one another. Equation (2.17) contains not only the angular frequency ω, but also the amplitudes a_1 and a_2. Substituting the single nonzero frequency found for this case (namely $\omega = \sqrt{2k/m}$) in Equation (2.17), we obtain a single relation between amplitudes: $a_2 = -a_1$. No further specification of the amplitudes is possible here because we have not specified the second initial condition—for example, the speed of the two masses—necessary for a complete determination of the motion. In terms of a_1, the motion of the two masses can be given as

motion of the (only) mode

$$u_1 = a_1 \sin \omega t, \qquad u_2 = a_2 \sin \omega t = -a_1 \sin \omega t = -u_1 \tag{2.18}$$

Let us understand the physical meaning of this.

The center of mass of the two springs can generally be written as

$$x_{\text{cm}} = \frac{m_1 x_1 + m_2 x_2}{m_1 + m_2} = \frac{x_1 + x_2}{2} = \frac{u_1 + u_2}{2} + \frac{x_{10} + x_{20}}{2}$$

or

$$x_{\text{cm}} = \frac{u_1 + u_2}{2} + x_{0\text{cm}}$$

where we have used the definition of u_1 and u_2 and the fact that $m_1 = m_2 = m$. $x_{0\text{cm}}$ is simply the initial position of the center of mass. But (2.18) implies that $u_1 + u_2 = 0$, so that x_{cm} maintains it initial value of $x_{0\text{cm}}$ for the entire motion: the system as a whole does not move. Let us choose the origin to be the center of mass, i.e., set $x_{0\text{cm}} = 0$. Then $x_{20} = -x_{10}$, and

$$x_2 = u_2 + x_{20} = -u_1 - x_{10} = -x_1$$

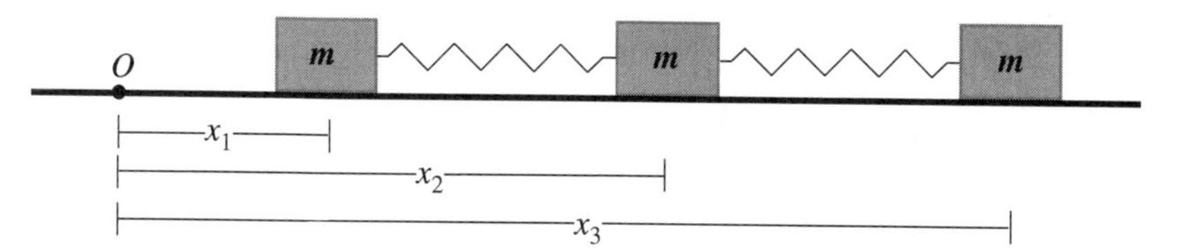

FIGURE 2.10. Three masses attached by two strings moving along a straight line on a frictionless horizontal surface.

It follows that the only nonzero mode of oscillation of the two masses corresponds to the case where they oscillate in unison but in opposite directions about the center of mass.

2.5.2 A System of Three Masses

To gain further insight into the motion of a system of masses and springs, let us examine three masses connected by two springs, as shown in Figure 2.10. The equations of motion for this system are

$$\begin{aligned} m\ddot{x}_1 &= k(x_2 - x_1 - L) \\ m\ddot{x}_2 &= -k(x_2 - x_1 - L) + k(x_3 - x_2 - L) \\ m\ddot{x}_3 &= -k(x_3 - x_2 - L) \end{aligned}$$

Note that the middle mass is pulled by two springs: one to the right and one to the left. As in the case of two masses, we reduce the x's to the u's, getting

$$\begin{aligned} m\ddot{u}_1 &= -ku_1 + ku_2 \\ m\ddot{u}_2 &= ku_1 - 2ku_2 + ku_3 \\ m\ddot{u}_3 &= ku_2 - ku_3 \end{aligned} \tag{2.19}$$

Furthermore, we assume that all masses move according to $u = a \sin \omega t$ with appropriate angular frequencies and amplitudes to be determined later. This turns (2.19) into

$$\begin{aligned} [(-m\omega^2 + k)a_1 - ka_2] \sin \omega t &= 0 \\ [-ka_1 + (-m\omega^2 + 2k)a_2 - ka_3] \sin \omega t &= 0 \\ [-ka_2 + (-m\omega^2 + k)a_3] \sin \omega t &= 0 \end{aligned}$$

Again, for this equation to hold for arbitrary t, we must have

$$\begin{pmatrix} k - m\omega^2 & -k & 0 \\ -k & 2k - m\omega^2 & -k \\ 0 & -k & k - m\omega^2 \end{pmatrix} \begin{pmatrix} a_1 \\ a_2 \\ a_3 \end{pmatrix} = \begin{pmatrix} 0 \\ 0 \\ 0 \end{pmatrix} \tag{2.20}$$

In order to obtain a nontrivial solution, we demand that

$$\det \begin{pmatrix} k - m\omega^2 & -k & 0 \\ -k & 2k - m\omega^2 & -k \\ 0 & -k & k - m\omega^2 \end{pmatrix} = 0$$

condition for the existence of solutions

or

$$m^3\omega^6 - 4km^2\omega^4 + 3k^2m\omega^2 = 0$$

Then, ignoring the $\omega = 0$ solution, we get

$$m^2\omega^4 - 4km\omega^2 + 3k^2 = 0$$

whose roots are $\omega_1 = \sqrt{k/m}$ and $\omega_2 = \sqrt{3k/m}$. Thus, *there are two normal modes for three masses connected by two springs.*

Substituting ω_1 in Equation (2.20), we obtain

$$\begin{pmatrix} 0 & -k & 0 \\ -k & k & -k \\ 0 & -k & 0 \end{pmatrix} \begin{pmatrix} a_1 \\ a_2 \\ a_3 \end{pmatrix} = \begin{pmatrix} 0 \\ 0 \\ 0 \end{pmatrix},$$

oscillation of the first mode

which yields $a_2 = 0$, and $a_3 = -a_1$. The displacements are, therefore,

$$u_1 = a_1 \sin \omega_1 t, \quad u_2 = 0, \quad u_3 = a_3 \sin \omega_1 t = -a_1 \sin \omega_1 t = -u_1 \tag{2.21}$$

As before, we assume that the center of mass does not move. Then Equation (2.21) describes a mode in which the middle mass is stationary, and the other two masses move in opposite directions with equal amplitudes. This situation is depicted in the vertical sequence of graphs on the left of Figure 2.11, where k, m, and a_1 are all set equal to 1. The top plot is u_1, the middle plot shows u_2 (which is zero, thus no displacement), and the bottom plot shows u_3, which is opposite to the direction of u_1.

Substitution of ω_2 in Equation (2.20) results in

$$\begin{pmatrix} -2k & -k & 0 \\ -k & -k & -k \\ 0 & -k & -2k \end{pmatrix} \begin{pmatrix} a_1 \\ a_2 \\ a_3 \end{pmatrix} = \begin{pmatrix} 0 \\ 0 \\ 0 \end{pmatrix}$$

oscillation of the second mode

which yields $a_2 = -2a_1 = -2a_3$, leading to the displacements

$$u_1 = u_3 = a_1 \sin \omega_2 t, \quad u_2 = -2a_1 \sin \omega_2 t \tag{2.22}$$

Equation (2.22) describes a mode in which the end masses are moving in unison in the same directions, while the middle mass is moving in the opposite direction with twice the amplitude, keeping the center of mass stationary. This motion is depicted in the vertical sequence of graphs on the right of Figure 2.11.

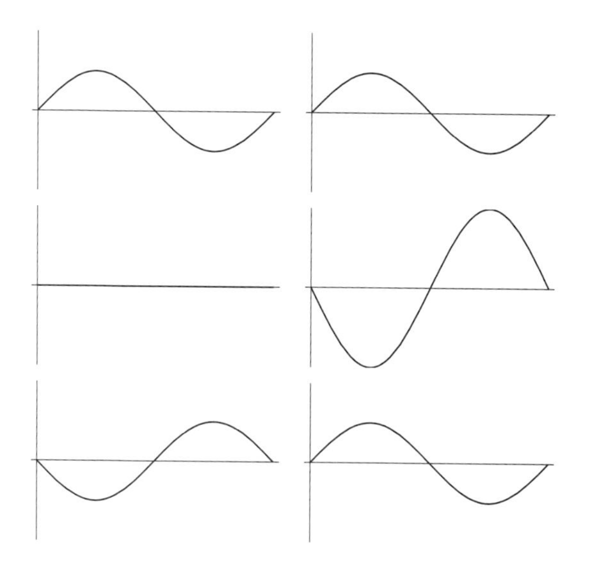

FIGURE 2.11. The two motions corresponding to the normal modes of three masses. The left column of graphs shows the motion corresponding to ω_1. The right column shows the motion corresponding to ω_2. In each case, the top plot shows u_1, the middle u_2, and the bottom u_3.

2.5.3 A System of Five Masses

The analytic study of the two cases of two and three masses has prepared us for the investigation of the general case. However, before doing so (in the next section), let us first solve the problem of five masses with the help of *Mathematica*. This will prepare us to attack the general case using the power of *Mathematica*.

Generalizing the three-mass case, the reader can show that the relevant matrix equation for five masses connected by four springs is

$$\begin{pmatrix} k - m\omega^2 & -k & 0 & 0 & 0 \\ -k & 2k - m\omega^2 & -k & 0 & 0 \\ 0 & -k & 2k - m\omega^2 & -k & 0 \\ 0 & 0 & -k & 2k - m\omega^2 & -k \\ 0 & 0 & 0 & -k & k - m\omega^2 \end{pmatrix} \begin{pmatrix} a_1 \\ a_2 \\ a_3 \\ a_4 \\ a_5 \end{pmatrix} = \begin{pmatrix} 0 \\ 0 \\ 0 \\ 0 \\ 0 \end{pmatrix}$$

Our task is to have *Mathematica* solve this equation for ω and the amplitudes.

First, we type in the matrix

```
In[1]:= m5={{k-m w^2,-k,0,0,0}, {-k,2 k-m w^2,-k,0,0},
        {0,-k,2 k-m w^2,-k,0},
        {0,0,-k,2 k-m w^2,-k}, {0,0,0,-k,k-m w^2}};
```

Recall that solutions come with → and { }.

Then we set its determinant equal to zero, solve the resulting equation for ω, and use `sol` to denote the solution:

```
In[2]:= sol=Solve[Det[m5]==0,w]
```

$$\left\{\{w\to 0\},\{w\to 0\},\left\{w\to -\frac{\sqrt{\frac{3k-\sqrt{5}k}{m}}}{\sqrt{2}}\right\},\left\{w\to \frac{\sqrt{\frac{3k-\sqrt{5}k}{m}}}{\sqrt{2}}\right\},\right.$$
$$\left\{w\to -\frac{\sqrt{\frac{5k-\sqrt{5}k}{m}}}{\sqrt{2}}\right\},\left\{w\to \frac{\sqrt{\frac{5k-\sqrt{5}k}{m}}}{\sqrt{2}}\right\},\left\{w\to -\frac{\sqrt{\frac{3k+\sqrt{5}k}{m}}}{\sqrt{2}}\right\},$$
$$\left.\left\{w\to \frac{\sqrt{\frac{3k+\sqrt{5}k}{m}}}{\sqrt{2}}\right\},\left\{w\to -\frac{\sqrt{\frac{5k+\sqrt{5}k}{m}}}{\sqrt{2}}\right\},\left\{w\to \frac{\sqrt{\frac{5k+\sqrt{5}k}{m}}}{\sqrt{2}}\right\}\right\}$$

In[3] shows one way of getting rid of → and { } to obtain the "naked" solution.

Next, we pick the positive frequencies and label them 1 through 4:

```
In[3]:= omeg1=Part[sol,4];w1=w/.omeg1;
        omeg2=Part[sol,6];w2=w/.omeg2;
        omeg3=Part[sol,8];w3=w/.omeg3;
        omeg4=Part[sol,10];w4=w/.omeg4;
```

Here `omeg1` is the fourth part of `sol`, i.e.,

$$\left\{w\to \frac{\sqrt{\frac{3k-\sqrt{5}k}{m}}}{\sqrt{2}}\right\}$$

and `w1=w/.omeg1` instructs *Mathematica* to evaluate `w` in `omeg1` and call it `w1`. The rest of `In[3]` evaluates `w2` through `w4`. So now we have all the nonzero frequencies.

Now we want to find the five amplitudes for each frequency. We construct the column matrix containing a_1 through a_5:

```
In[4]:= A={a1,a2,a3,a4,a5};
```

and multiply `A` on the left by `m5`:

```
In[5]:= mDotA=m5.A

Out[5]:=
```

$$\{-a_2k+a_1(k-mw^2),-a_1k-a_3k+a_2(2k-mw^2),-a_2k-a_4k$$
$$+a_3(2k-mw^2),-a_3k-a_5k+a_4(2k-mw^2),-a_4k+a_5(k-mw^2)\}$$

This is the column vector whose components are set equal to zero and subsequently solved for a_2 through a_5 in terms of a_1. So we set up the following five expressions:

```
In[6]:= lhs1=Part[mDotA,1]/.w->w1;
        lhs2=Part[mDotA,2]/.w->w1;
        lhs3=Part[mDotA,3]/.w->w1;
        lhs4=Part[mDotA,4]/.w->w1;
        lhs5=Part[mDotA,5]/.w->w1;
```

The first line defines `lhs1` to be the first part of `mDotA` with `w1` substituted for `w`; similarly for the other expressions. The input

```
In[7]:= solution=Solve[{lhs1==0,lhs2==0,lhs3==0,lhs4==0,
        lhs5==0},{a2,a3,a4,a5}]
```

produces

$$\left\{\left\{a5 \to -\frac{1}{4}\left(-1+\sqrt{5}\right)\left(1+\sqrt{5}\right)a1, a4 \to -\frac{1}{2}\left(-1+\sqrt{5}\right)a1,\right.\right.$$
$$\left.\left. a3 \to 0, a2 \to \frac{1}{2}\left(-1+\sqrt{5}\right)a1\right\}\right\}$$

Now we have to untangle the amplitudes from the curly brackets and arrows. This is done as follows:

In[8] shows another way of getting rid of → and { }.

```
In[8]:= a2=a2/.solution;a2=First[a2];
        a3=a3/.solution;a3=First[a3];
        a4=a4/.solution;a4=First[a4];
        a5=a5/.solution;a5=First[a5];
```

The first statement of each line extracts an amplitude from the `solution`; but it is in the form of a list (it still has curly brackets around it). The second statement then defines the new amplitude to be the first—and only—element of this list.

With the amplitudes corresponding to ω_1 determined, we can plot the five displacements. Again, we set k, m, and a_1 equal to 1. The first mode is shown in Figure 2.12 on the left. In this mode the end masses oscillate in opposite directions with the same amplitude, and the next two masses also oscillate in opposite directions with the same amplitude that is smaller than those of the end masses. The middle mass has no motion relative to the center of mass, which happens to be at rest.

The remaining modes can also be interpreted. In the second mode, the end masses move in phase with the same amplitude. The second and fourth masses also move in phase, but with half the amplitude of the end masses. To keep the center of mass stationary, the middle mass helps the second and fourth masses by oscillating in their direction with an amplitude equal to that of the end masses. The third mode is similar to the first mode, except that the end masses are oscillating with the smaller amplitude. The last mode is similar to the second mode, but the amplitude of the end masses is now small, causing the middle mass to oscillate in phase with them.

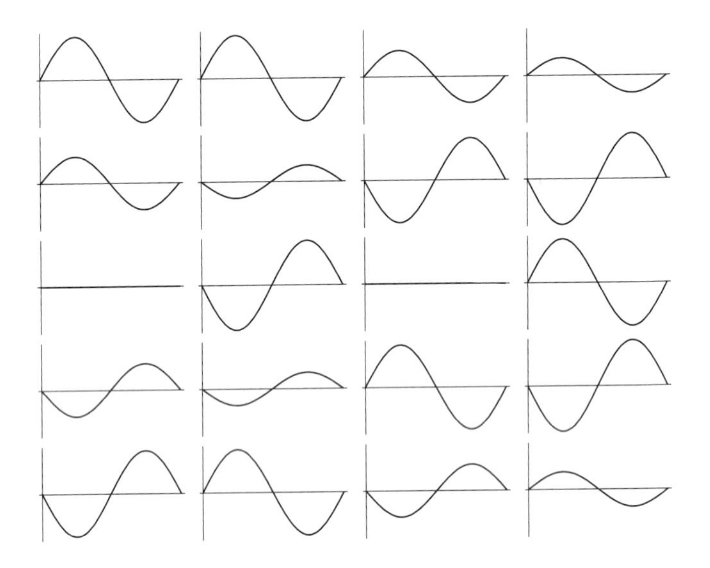

FIGURE 2.12. The four motions corresponding to the four normal modes of five masses. Each column of graphs shows the motion of the five masses corresponding to a particular frequency.

2.6 Normal Modes of a System of n Masses

We have a pretty good idea of how normal modes work—at least for up to five masses. We now want to solve the problem for n masses. In the process of solving this problem, we learn a great deal about how *Mathematica* deals with matrices and vectors.

The first—and biggest—challenge is to construct the $n \times n$ matrix that generalizes the 5×5 matrix of the last section. We saw that only two or three elements of any given row are nonzero. Therefore, there is no single "formula" for elements of the matrix, or even its rows. In fact, the values of the row elements are *conditional* upon their locations. *Mathematica* handles conditional cases using logical variables and operators. So, let us discuss this briefly.

Suppose we want to define a function $g(t)$ that is $+1$ if t is positive and 0 if t is negative. As the definition suggests, we have to incorporate *conditions* in the definition of the function. This code accomplishes the task:

how to construct a discontinuous function

```
In[1]:= f[t_]:=If[t>0,+1,0]
```

The following are some of the logical expressions used in *Mathematica*:

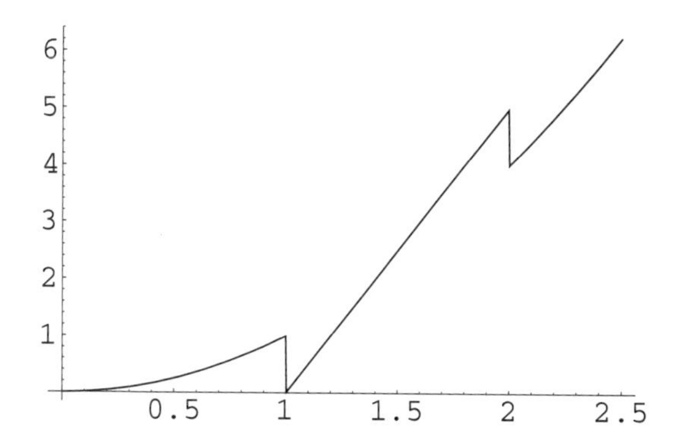

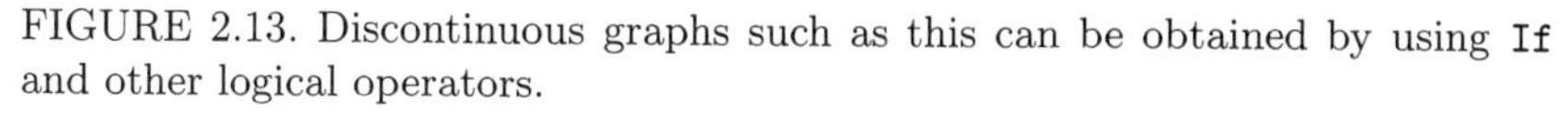

FIGURE 2.13. Discontinuous graphs such as this can be obtained by using `If` and other logical operators.

`If[test,e1,e2]`	give $e1$ if *test* succeeds, and $e2$ if it fails
`||`	the logical operator `or`
`&&`	the logical operator `and`
`!`	the logical operator `not`

Note that `e1` and `e2` could be mathematical expressions. For example,

```
In[1]:= g[t_]:=If[t>1 && t<2,5(t-1),t^2];
        Plot[g[t],{t,0,2.5},PlotRange->All]
```

will produce Figure 2.13. The discontinuities at $t = 1$ and $t = 2$ are caused by a change in the formula of the function.

So, how do we use conditional statements to construct our $n \times n$ matrix? Let us start with

conditional statements used to construct the normal-mode matrix

```
In[1]:= b[i_,j_]:=If[(j<i-1||j>i+1),0,-k];
        bDiag[i_,n_]:=If[(i!=1)&&(i!=n),2k-m w^2,k-m w^2];
```

What does this input mean? The first line defines the off-diagonal elements of the matrix. It instructs *Mathematica* to put zero for the jth element of the ith row if $j < i - 1$ or $j > i + 1$. Everywhere else, *including at the ith location*, *Mathematica* puts $-k$. This is fine because the next statement overrides the unwanted result. The second line of the input defines the diagonal elements. It tells *Mathematica* to substitute $2k - m\omega^2$ when $i \neq 1$ and $i \neq n$, and $k - m\omega^2$ otherwise.

The two statements of `In[1]` define certain functions, but they are not connected and by themselves do not construct the elements of a matrix. However, the ijth element of our matrix connects them:

elements of the matrix of normal mode constructed

```
In[2]:= c[i_,j_,n_]:=If[i!=j,b[i,j],bDiag[i,n]]
```

where `c[i,j,n]` stands for the ijth element of an $n \times n$ matrix. Notice that `c[i,j,n]` uses `b[i,j]` when $i \neq j$ and `bDiag[i,n]` otherwise (i.e., when $i = j$). Having constructed the elements, we can build the matrix itself using `Table`:

the matrix itself constructed

```
In[3]:= matr[n_]:=Table[c[i,j,n],{i,1,n},{j,1,n}]
```

To test our code, we type in

```
In[4]:= matr[5]//MatrixForm
```

and *Mathematica* produces the output

$$\begin{pmatrix} k - m\omega^2 & -k & 0 & 0 & 0 \\ -k & 2k - m\omega^2 & -k & 0 & 0 \\ 0 & -k & 2k - m\omega^2 & -k & 0 \\ 0 & 0 & -k & 2k - m\omega^2 & -k \\ 0 & 0 & 0 & -k & k - m\omega^2 \end{pmatrix}$$

which is the matrix we encountered in our discussion of the normal modes of five masses.

Since the roots of polynomials of degree 5 and above cannot be found using radicals, we have to resort to numerical solutions; and for this, we need to assign numerical values to all the parameters of the problem. So we set k and m equal to 1 for convenience:

```
In[5]:= k=1; m=1;
```

solving for normal frequencies

and find the roots of the determinant of the matrix numerically:

```
In[6]:= sol[n_]:=NSolve[Det[matr[n]]==0,w]
```

This is possible because the determinant of `matr[n]` is always a polynomial in ω.

Next, we want to extract the frequencies from `sol[n]`, i.e., get rid of the arrows and curly brackets. This is done as before:

extracting the normal frequencies from solutions

```
In[7]:= omeg[m_,n_]:=Part[sol[n],n+m+1];
        w[m_,n_]:=w/.omeg[m,n]
```

Note that the mth frequency is the $n+m$th `Part` of `sol[n]`. This is because *Mathematica* orders the roots of the determinant in ascending order starting with the negative solutions. This ordering results in the nth and $n+1$st frequencies being zero. Therefore, the frequencies defined in `In[7]` include only the positive frequencies—a total of $n-1$—in ascending order.

We now define the column vector of amplitudes and multiply it on the left by our matrix:

constructing the vector of amplitudes

```
In[8]:= A[n_]:=Array[a,n];mDotA[n_]:=matr[n].A[n]
```

where the command `Array[a,n]` produces a vector with elements `a[1]` through `a[n]`. The other function, `mDotA[n]`, is an n-dimensional column vector whose elements are to be set equal to zero and the resulting equations solved to yield the amplitudes. The left-hand sides of these equations are obtained thus:

constructing the left-hand side of amplitude equations

```
In[9]:= lhs[m_,j_,n_]:=Part[mDotA[n],m]/.w->w[j,n]
```

Earlier, we listed the equations and the sought-after amplitudes explicitly, as in `In[7]` of the five-mass system. Since n is arbitrary, we cannot have an explicit listing of the amplitudes here. We have to replace the explicit list with a `Table`. The following code will do the job:

solving for $n - 1$ amplitudes in terms of `a[1]`

```
In[10]:= solution[j_,n_]:=Chop[Simplify[NSolve[Table
         [lhs[i,j,n]==0,{i,1,n-1}],Table[a[i],{i,2,n}]]]]
```

Let us explain the statement above. `Chop` eliminates any insignificant small number left over from numerical calculations. The first `Table` lists the first $n-1$ equations involving the amplitudes. Although there are n equations, they are not all independent. After all, we are interested in $n-1$ amplitudes given in terms of the first; and for this we do not need all n equations. We could get away with this "overspecification" in the case of three and five masses, because we listed the amplitudes explicitly. In this general case, we receive an error message if we try to solve all n equations. The second `Table` lists `a[2]` through `a[n]`, and `NSolve` solves this set of $n-1$ equations in $n-1$ unknowns.

use of **Chop**

The remaining task is to extract the amplitudes from `In[10]` and construct the solutions. The following statement extracts the amplitudes:

extracting amplitudes from solutions

```
In[11]:= tbl[j_,n_]:=Table[a[i]/.solution[j,n],{i,2,n}];
        a[i_,j_,n_]:=Part[tbl[j,n],i-1];
        amp[nAmp_,nMode_,nSpring_]:=First[a[nAmp,
        nMode,nSpring]]
```

The first line makes a list of the amplitudes obtained in `solution[j,n]`. The second line takes the $(i-1)$st `Part` of the resulting table and calls it `a[i,j,n]` with $i \geq 2$. The last line gets rid of the curly brackets around `a[i,j,n]`. Note that in the last line we changed the dummy variables to a more suggestive notation.

The solutions can now be written down:

equation of each mode of oscillation

```
In[12]:= u[nAmp_,nMode_,nSpring_,t_]:=
          amp[nAmp,nMode,nSpring] Sin[w[nMode,nSpring] t]
        u[1,nMode_,nSpring_,t_]:=
           a[1] Sin[w[nMode,nSpring] t]
```

If desired, we can plot the displacements of various masses as a function of time. For example, to plot the displacement of the second mass in a collection of seven masses corresponding to the third frequency, we type in

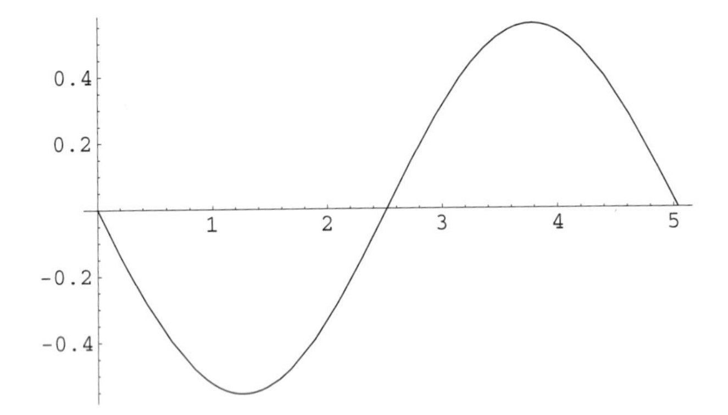

FIGURE 2.14. The displacement corresponding to the third frequency of the second mass in a collection of seven masses.

```
In[13]:= Plot[u[2,3,7,t],{t,0,2 Pi/w[3,7]}]
```

and obtain Figure 2.14.

2.7 Problems

Problem 2.1. Using *Mathematica*,
(a) find the dot product of $\mathbf{a} = \langle 1, 1, -1 \rangle$ with the cross product of the two vectors $\mathbf{b} = \langle 2, -1, 3 \rangle$ and $\mathbf{c} = \langle 1, -1, 2 \rangle$.
(b) Find the cross product of $\mathbf{a} = \langle 1, 1, -1 \rangle$ with the cross product of the two vectors $\mathbf{b} = \langle 2, -1, 3 \rangle$ and $\mathbf{c} = \langle 1, -1, 2 \rangle$.

MM, Problem 1.14

Problem 2.2. Let $\mathbf{a} = \langle a_x, a_y, a_z \rangle$, $\mathbf{b} = \langle b_x, b_y, b_z \rangle$, and $\mathbf{c} = \langle c_x, c_y, c_z \rangle$. Using *Mathematica*,
(a) find the double cross product $\mathbf{a} \times (\mathbf{b} \times \mathbf{c})$ in terms of the components of the three vectors. Check your answer against the so-called *bac cab rule*.
(b) Verify the following two identities:

$$\mathbf{a} \cdot (\mathbf{b} \times \mathbf{c}) = \mathbf{c} \cdot (\mathbf{a} \times \mathbf{b}) = \mathbf{b} \cdot (\mathbf{c} \times \mathbf{a})$$
$$(\mathbf{a} \times \mathbf{b}) \cdot (\mathbf{a} \times \mathbf{b}) = |\mathbf{a}|^2 |\mathbf{b}|^2 - (\mathbf{a} \cdot \mathbf{b})^2$$

Problem 2.3. Using Equation (2.7) and an n larger than 60, try to achieve a value for the three-dimensional Madelung constant accurate to two decimal places (i.e., 1.74).

Problem 2.4. In Equation (2.8) replace n for ∞ in both sums. See how large n should be for the double sum to yield the Madelung constant given in the second line of Equation (2.8).

Problem 2.5. Using Equation (2.14), "tube" a spiral whose equation in polar coordinates is $r = 0.01\theta$. Try different values for the radius of the tube. Make sure the range of θ is large enough for you to see a few turns of the spiral in the parametric plot of the tube. Hint: Write the single polar equation as two Cartesian equations.

Problem 2.6. Refer to your calculus book, and using Equation (2.14), "tube" some of the nicer-looking polar curves you find there.

Problem 2.7. Using *Mathematica*, "prove" that if you exchange the first and second rows of a general 3×3 matrix, its determinant changes sign.

Problem 2.8. Using *Mathematica*, "prove" that if two rows or two columns of a general 4×4 matrix are equal, its determinant vanishes.

Problem 2.9. Solve the following linear systems of equations using matrices in *Mathematica.*

$$\text{(a)} \quad \begin{aligned} x + y - 2z &= 1 \\ 3x - y + z &= 2 \\ x + 4y - 3z &= 0 \end{aligned} \qquad \text{(b)} \quad \begin{aligned} 2x - y &= 3 \\ x + y - z &= -1 \\ -x + 2y + 2z &= 2 \end{aligned}$$

$$\text{(c)} \quad \begin{aligned} x + y - z + 2w &= 1 \\ 2x - y + z - w &= -1 \\ x + 2y - z + w &= 2 \\ x + y - 2w &= -2 \end{aligned} \qquad \text{(d)} \quad \begin{aligned} x + 2y - 3z + 2w &= 1 \\ x - y - z + 3w &= -1 \\ 2x + y - z + 2w &= 3 \\ 4x + 3y - 2z + w &= 0 \end{aligned}$$

Problem 2.10. Four 1-kg masses are linearly connected by three springs with spring constant $k = 0.02$. Find the normal frequencies and plot the motion of each of the four masses corresponding to any two normal frequencies.

Problem 2.11. Six 1-kg masses are linearly connected by four springs with spring constant $k = 50$. Find the normal frequencies and plot the motion of each of the five masses corresponding to the lowest normal frequency.

Problem 2.12. Eight 1-kg masses are linearly connected by five springs with spring constant $k = 2$. Find the normal frequencies, and plot the motion of each of the six masses corresponding to the two lowest normal frequencies.

3
Integration

The significance of the concept of integration as one of the most fundamental processes of mathematical physics is no doubt familiar to the reader. In fact, it is no exaggeration to claim that modern mathematics and physics started with this concept. Generally speaking, physical laws are given in local form while their application to the real world requires a departure from locality. For instance, the universal law of gravity is given in terms of point particles, actual *mathematical points*, and the law, written in the language of mathematics, assumes that. In real physical situations, however, we never deal with a mathematical point. Usually, we *approximate* the objects under consideration as points, as in the case of the gravitational force between the Earth and the Sun. Whether such an approximation is good depends on the properties of the objects and the parameters of the law. In the example of gravity, on the sizes of the Earth and the Sun as compared to the distance between them. On the other hand, the precise motion of a satellite circling the Earth requires more than approximating the Earth as a point; all the bumps and grooves of the Earth's surface will affect the satellite's motion.

Physical laws are given for mathematical points but applied to extended objects.

The application of physical laws—given for mathematical points—to extended everyday objects requires integration, the subject of this chapter. We shall consider various examples from selected branches of physics.

3.1 Integration in *Mathematica*

Mathematica has a wide variety of integration techniques, but these techniques are not conveniently available to the user. One has to use discretion in one's choice of the method of integration. For example, although *Mathematica* can handle multiple integrals, it is often good practice to have it perform some of the integrals separately and then feed the results into the remaining integrals. It may also be expedient—when doing a definite integral—to break up the process into two parts: first have *Mathematica* do the *indefinite* integral, then tell it to substitute the limits. In short, integration in *Mathematica* is an art rather than a science, and we shall see numerous examples of this in the following.

Integration in *Mathematica* is an art, not a science!

There are two commands for integration in *Mathematica*, `Integrate` and `NIntegrate`, with the second one doing numerical integration only. For the first, the analytical one, we can either specify the limits of integration (definite integral), or ignore it (indefinite integral). Here is the syntax for the analytical integration:

analytical and numerical integration in *Mathematica*

`Integrate[f[t],t]`	give the indefinite integral of f
`Integrate[f[t],{t,a,b}]`	evaluate $\int_a^b f(t)\,dt$
`Integrate[f[t,u],{t,a,b},{u,c,d}]`	evaluate $\int_a^b dt \int_c^d du f(t,u)$

In the double integral, the u integration is done first and then the t integration. Furthermore, c and d could be functions of t.

3.1.1 *The Simple Example of a Pendulum*

Let us start with a simple example from mechanics, where it is shown that the period of a pendulum is given in terms of a definite integral as follows:

MM, pp. 271–273

$$T = 4\sqrt{\frac{l}{g}}\int_0^{\pi/2} \frac{du}{\sqrt{1-\sin^2(\theta_m/2)\sin^2 u}} \tag{3.1}$$

where l is the length of the pendulum, g is the gravitational acceleration at the location of the pendulum, and θ_m is the maximum angle—from vertical—reached by the pendulum. Note that for small maximum angle ($\theta_m \approx 0$) the integrand reduces to 1 and the period becomes $T = 2\pi\sqrt{l/g}$, a familiar result stating in particular that the period is independent of the (small) angle—a result that was familiar to Galileo. To tell *Mathematica* to do the integral of Equation (3.1), type in

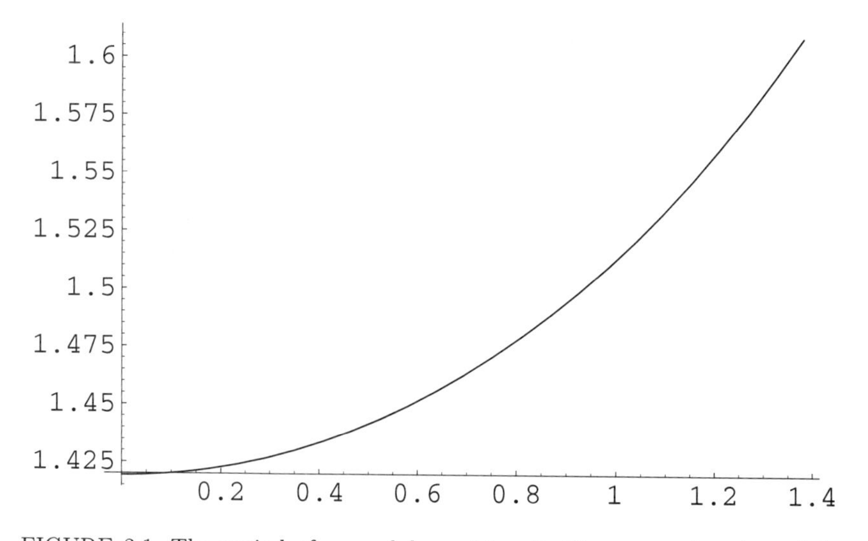

FIGURE 3.1. The period of a pendulum of length 50 cm as a function of the maximum angle (in radians).

```
In[1]:= period[theta_,length_,g_]:=4 Sqrt[length/g]
        Integrate[1/Sqrt[1-Sin[theta/2]^2 Sin[u]^2],
        {u,0,Pi/2}]
```

With the function `period` so defined, we can tell *Mathematica* to plot it for us. The command

```
In[2]:= Plot[period[t,0.5,9.8],{t,0,1.4}]
```

will produce Figure 3.1 for $l = 0.5$ meter and $g = 9.8$ m/s^2. Notice how we used `t` instead of `theta` as the first argument of `period`. Once the function is defined, any symbol can be used as its argument.

3.1.2 Shortcomings of Numerical Integration in Mathematica

We are not always as lucky as the example above may suggest! In many instances *Mathematica* does a poor job of integrating a function. For example, consider the innocent-looking function e^{-x^2}, whose integral over the entire real line is $\sqrt{\pi} = 1.772454$. If you integrate this function over the interval $(-5, 5)$, and ask for an answer to seven significant figures,

MM, pp. 96–97

```
In[3]:= N[NIntegrate[Exp[-x^2],{x,-5,5}],7]
```

you will get 1.772454. However, if you try to make the integral more accurate by extending the region of integration, you will succeed at the beginning, but when you reach large values of the integration limits, you will get surprising results! For instance, upon the input

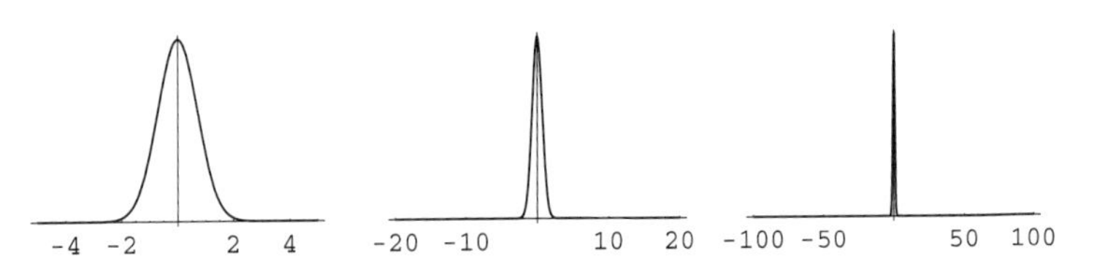

FIGURE 3.2. The "smooth" function e^{-x^2} (left) in the interval $(-5, 5)$ becomes less smooth in the interval $(-20, 20)$ in the middle, until it turns into a very sharp spike in the interval $(-100, 100)$.

```
In[4]:= NIntegrate[Exp[-x^2],{x,-500,500}]
```

Mathematica will yield 0.88631 after complaining about slow convergence of the numerical integration. What is the reason? As the output of `In[3]` indicates, most of the contribution to the integral comes from values between -5 and $+5$. When *Mathematica* evaluates an integral numerically over an interval of the real line, it takes sample points of the interval, evaluates the function at those points, and sums up the contributions. In the interval $(-500, 500)$, which is 100 times the size of the interval $(-5, 5)$, in which the function lends its entire contribution, *Mathematica* will miss most of the latter interval. The result is that it adds up a lot of small numbers. In fact, the larger you make the interval, the smaller the result will be! For the interval $(-800, 800)$, you will get 8.56868×10^{-17}, and for $(-1000, 1000)$, the result will be 1.34946×10^{-26}.

Figure 3.2 gives an "explanation" for all this. While the function e^{-x^2} is smooth in the interval $(-5, 5)$, it turns into a sharp spike in the interval $(-100, 100)$. In numerically integrating a function, *Mathematica* has only a finite sequence of values for the function. Therefore, it has to make certain assumptions about the behavior of the function. In particular, it assumes that the function is smooth over the interval of integration; i.e., that the function does not change abruptly. For e^{-x^2} and the interval $(-5, 5)$, this assumption is valid, but—as Figure 3.2 clearly indicates—once you reach intervals larger than $(-100, 100)$, the assumption no longer holds.

One can improve the result by forcing *Mathematica* to include an interval in the neighborhood of the peak. This is done by inserting the end values of the interval of interest between the two limits of integration. For example,

```
In[5]:= NIntegrate[Exp[-x^2],{x,-500,-3,3,500}]
```

will return 1.77241.

The following are the various numerical integration commands in *Mathematica*:

`NIntegrate[f[t],{t,a,b}]`	evaluate $\int_a^b f(t)\,dt$ numerically
`NIntegrate[f[t,u],{t,a,b},{u,c,d}]`	evaluate $\int_a^b dt \int_c^d du\, f(t,u)$ numerically
`NIntegrate[f[t],{t,a,c ... ,d,b}]`	evaluate $\int_a^b f(t)\,dt$ numerically making sure to include points $c, \ldots, d$

The first one is the simplest kind of numerical integration command; the second is a double integral—which could be extended to any number of dimensions—in which the u integration is done first and then the t integration. The limits of u could be functions of t. The third inserts some strategic points in the interval of integration, forcing *Mathematica* to sample points in the inserted intervals.

3.1.3 Other Intricacies of Integration in Mathematica

To learn more about the intricacies of integration in *Mathematica*, let us calculate the area of a circle of radius a in Cartesian coordinates. This is a double integral of the form

$$\int_{-a}^{a} dt \int_{-\sqrt{a^2-t^2}}^{\sqrt{a^2-t^2}} du$$

We first do this numerically for a circle of unit radius. So, we type in

```
In[1]:= NIntegrate[1,{t,-1,1},{u,-Sqrt[1-t^2],
        Sqrt[1-t^2]}]//Timing
```

and get {1.44Second, 3.14159}, indicating that it took *Mathematica* 1.44 seconds to obtain the value of π—the area—to six significant figures.[1] Note the use of `//Timing`, which tells *Mathematica* to display the time spent on the calculation.

use of **//Timing**

Now let us do the same calculation analytically by typing in

```
In[2]:= Integrate[1,{t,-a,a},{u,-Sqrt[a^2-t^2],
        Sqrt[a^2-t^2]}]//Timing
```

[1]Actually the *displayed* result has six significant figures of accuracy. The internal calculation is more accurate.

and waiting, ... , and waiting! After 15 minutes we give up[2] and try to do the integral a different way. As indicated earlier, it is a good idea to break up the two integrations. We can easily do the trivial u integration, and have *Mathematica* do the single integral

$$2\int_{-a}^{a} \sqrt{a^2 - t^2}\, dt$$

So, we type in

```
In[3]:= Integrate[Sqrt[a^2-t^2],t^2],{t,-a,a}]//Timing
```

and wait, ... , and wait! Another 15 minutes go by, and *Mathematica* is still "thinking." We abort the calculation, and try to find yet another way of calculating the integral.

Now we tell *Mathematica* to find the *indefinite* integral:

```
In[4]:= g=Integrate[Sqrt[a^2-t^2],t^2],t]//Timing
```

and get the answer in a fraction of a second:

$$\left\{0.26 Second, t\sqrt{a^2 - t^2} - a^2 \,\text{ArcTan}\left[\frac{t\sqrt{a^2 - t^2}}{-a^2 + t^2}\right]\right\}$$

We try to evaluate this expression at the limits of integration. So, we type in

```
In[5]:= (g/.t->a)-(g/.t->-a)
```

But *Mathematica* cannot handle the implicit infinity in the argument of `ArcTan` and so returns `Indeterminate` as the output.

The inclusion of the parentheses in `In[5]` is extremely important. Let us illustrate why. If you enter `f=z^2` first and then `f/.z->b-f/.z->a`, you will get $(a - b^2)^2$ as the output! Enclosing the the terms in parentheses gives the intended answer, $b^2 - a^2$.

The output `Indeterminate` may seem reasonable, as both the numerator and the denominator of the argument of the `ArcTan` are zero. However, even if you simplify the argument to $-t/\sqrt{a^2 - t^2}$—something *Mathematica* does not do automatically—and then type in `g/.t->a`, the output is still `Indeterminate`, which is wrong! Although *Mathematica* knows that `ArcTan[∞]` is $\pi/2$ and that $t/\sqrt{a^2 - t^2}$ is infinite at $t = a$, it cannot make the connection.

taking limits of functions

How *do* we make *Mathematica* evaluate g at the two limits of integration? Let's try using `Limit[f[t],t->a]`, which is equivalent to $\lim_{t\to a} f(t)$. Typing in

[2] The calculation was done on an *iMac* using the front end of a remote kernel.

```
In[6]:= Limit[g,t->a]
```

yields $-a^2\pi/2$, which, except for the sign, is what we are looking for. The problem is that the lower-limit calculation, i.e., `Limit[g,t->-a]`, yields exactly the same result; so the area is calculated to be zero! The reason for the dilemma is not making a distinction between the two different approaches to a. When we take the limit to a, we are approaching it *from below*, because we want to remain in the range of integration. On the other hand, when we take the limit to $-a$, we are approaching it *from above*, for the same reason. So try

Directional limits are sometimes useful in evaluating integrals.

```
In[7]:= Limit[g,t->a,Direction->1]
```

and obtain $a^2\pi/2$. Similarly,

```
In[8]:= Limit[g,t->-a,Direction->-1]
```

yields $-a^2\pi/2$, which together with the output of `In[7]` gives the area of the circle.

Here are the appropriate commands for various limits.

`Limit[f[t],t->a]`	find the limit of $f(t)$ as t approaches a
`Limit[f[t],t->a,Direction->1]`	find the limit as t approaches a from below
`Limit[f[t],t->a,Direction->-1]`	find the limit as t approaches a from above

The signs in the directional limits may be confusing, but if you think about approaching $+1$ and -1 from the origin, they make sense.

3.2 Integration in Mechanics

The master equation in mechanics is of course Newton's second law of motion. This is a differential equation and, as such, is most appropriately discussed in a treatment of that subject. Nevertheless, there are some problems that could be solved by integration. We treat some such problems in one dimension in which the applied force is a function of position only.

3.2.1 Position-Dependent Forces

In this subsection, we apply the second law of motion to a particle under the influence of a force that depends only on the position of the particle.

Let x be the coordinate of a particle of mass m, and $f(x)$ the force acting on it. The equation of motion is

$$m\frac{d^2x}{dt^2} = f(x) \quad \text{or} \quad \frac{dv}{dt} = \frac{f(x)}{m}$$

where $v = dx/dt$ is the speed of the particle. Now we use the common trick:

$$\frac{dv}{dt} = \frac{dv}{dx}\frac{dx}{dt} = v\frac{dv}{dx}$$

and rewrite the equation of motion as

$$v\frac{dv}{dx} = \frac{1}{m}f(x) \Rightarrow v\,dv = \frac{1}{m}f(x)\,dx$$

With x_0 and v_0 as initial position and speed, we can integrate the last equation:[3]

MM, pp. 82 and 100

$$\int_{v_0}^{v} u\,du = \frac{1}{m}\int_{x_0}^{x} f(s)\,ds \quad \text{or} \quad \tfrac{1}{2}v^2 - \tfrac{1}{2}v_0^2 = \frac{1}{m}\int_{x_0}^{x} f(s)\,ds$$

We now have a new equation to integrate:

$$v^2 = v_0^2 + \frac{2}{m}\int_{x_0}^{x} f(s)\,ds \Rightarrow \frac{dx}{dt} = \left(v_0^2 + \frac{2}{m}\int_{x_0}^{x} f(s)\,ds\right)^{1/2}$$

or

$$\frac{dx}{\sqrt{v_0^2 + \frac{2}{m}\int_{x_0}^{x} f(s)\,ds}} = dt \quad \text{or} \quad \int_{x_0}^{x} \frac{dx}{\sqrt{v_0^2 + \frac{2}{m}\int_{x_0}^{x} f(s)\,ds}} = t \tag{3.2}$$

Equation (3.2) gives the solution of the equation of motion *implicitly*; it gives t as a function of x. Therefore, one has to "solve" the equation for x as a function of t. It may be impossible to solve such an equation *analytically*, but the *numerical* solution may not be out of reach.

To start the process of solving Equation (3.2), let us define two functions in *Mathematica*:

```
In[1]:= F[x_,x0_]:=Integrate[f[s],{s,x0,x}];
        g[x_,x0_,v0_]:=Integrate[1/Sqrt[v0^2 +F[u_,x0_]],
        {u,x0,x}];
```

where, for simplicity, we have set $m = 2$. The solution then consists of solving the equation $g(x, x_0, v_0) = t$ for given values of x_0 and v_0 and various values of t. First, to get some insight, we let $x_0 = 1$, let $v_0 = 1$, and

[3]We follow the important advice that variables of integration are to be different from the symbols used for limits of integration. (See *MM*, pp. 82 and 100.)

try a value for t equal to 1. We have to know the force function, of course. So let $f(s) = s$.

Chapter 1 gave a brief description of methods for solving equations. Since `Solve` and `NSolve` work only for polynomials, we have no choice but to use `FindRoot`. So, let us type in

```
In[2]:= f[s_]:=s; FindRoot[g[x,1,1]==1,{x,1}]
```

Out[2]:= $\{x \to 2.34603\}$

The solution is in the form of a *rule*, not a number. However, if we type

```
In[3]:= x/.%
```

we get the number 2.34603—without a substitution rule—as the output. Because % always refers to the previous expression, we can give a name to the expression in `In[2]` and refer to that name in `In[3]`. Specifically, rewrite `In[2]` as

```
In[4]:= f[s_]:=s; y=FindRoot[g[x,1,1]==1,{x,1}];
        z=x/.y; z
```

and get the output

Out[4]:= 2.34603

Now all we have to do is to come up with a way of repeating the procedure above for many values of t. First we have to generate such values for t. In mathematics, we would label these as t_1, t_2, t_3, ... ; in *Mathematica* we label them as `t[1]`, `t[2]`, `t[3]`, ... There is a command in *Mathematica* that is particularly suited for this process; it is called `Do`. For our present purpose, the syntax goes as follows:

using **Do** to find values for the independent variable

```
In[5]:= t[0]=0; deltat[T_,n_]:= Do[t[i]=t[i-1]+T/n,{i,n}]
```

The first statement sets the initial time t_0 equal to zero. In the second statement, we have defined a function `deltat` of two variables n and T. The first variable gives the "final" time (or the duration of the observation of motion), and the second variable is the number of divisions of the time interval. The right-hand side of the second statement takes the previous value `t[i-1]` and adds an *increment* T/n to it to arrive at the new value `t[i]`. This calculation is done n times, as indicated by the list `{i,n}`. As an example, the input

```
In[6]:= deltat[5,10]; Table[t[i],{i,0,10}]
```

yields the following array as an output:

Out[6]:= $\{0, 0.5, 1., 1.5, 2., 2.5, 3., 3.5, 4., 4.5, 5.\}$

We have thus far learned how to divide the time values—the horizontal axis in a plot—and how to solve for x for a given value of t. We now have to combine these procedures and generate the x-values—the vertical axis in a plot. For this, we use the loop command `For`. The syntax of some of these repetition commands are given below:

`Do[expr,{i,n}]`	evaluate `expr` with i running from 1 to n
`Do[expr,{i,m,n,di}]`	evaluate `expr` with i running from m to n in steps of di
`For[start,test,incr,body]`	evaluate `start`, then repetitively evaluate `body` and `incr` until `test` fails
`Table[f[i],{i,m,n}]`	make a list of the values of f with i running from m to n

The argument `start` in `For` usually has a counting index i whose initial value is given in `start` and is increased according to the rule in `incr`. This rule is typically written as `i=i+di`, where `di` is the step size (or increment) for `i`. When `i++` is used as `incr`, the value of i is increased by one unit in each evaluation; thus `i++` is equivalent to `i=i+1`. The `test` is usually of the form $i \leq n$, whereby the evaluation is repeated until $i = n$ at which time the evaluation stops. The `body` is the heart of the calculation, and it can involve several commands separated by colons.

We are now ready to write down the commands that evaluate the implicit function of (3.2) for various values of t. It is

using **For** to calculate values of dependent variable for given values of independent variable

```
In[7]:= displ[n_,x0_,v0_]:= For[i=1;t[0]=0;x[0]=x0,i<=n,
        i++, y=FindRoot[g[x,x0,v0]==t[i],{x,x[i-1]},
        MaxIterations->50]; x[i]=x/.y;]
```

A number of remarks are in order. First, the argument `start` in `In[7]` introduces i as the counting index and initializes it to 1. Similarly, t and x are initialized to 0 and x_0, respectively. Second, the `test` in this case is $i \leq n$, written as `i<=n`. Third, we have used x_{i-1} as the starting value in `FindRoot`. This is reasonable, because we expect x_i—the solution x to the equation—to be close to x_{i-1}. So the latter should be a good estimate for the next value of x. Finally, because of the shortcomings of `FindRoot`, we have introduced the option `MaxIterations->50` to allow for more iteration than is allowed by *Mathematica*'s default value of 15.

using the option **MaxIterations**

So far we have suppressed outputs and only defined functions. For *Mathematica* to perform calculations, we need to put in some actual functions and numbers. Let us take $f(s) = 19.6$ corresponding to the uniformly accelerated free-fall motion on earth (recall that $m = 2$). We study the motion for

5 time units (5 seconds) and divide the time interval into 20 subintervals. The syntax for all of this is

```
In[8]:= f[s_]:=19.6; deltat[5,20]; displ[20,0.001,0.001];
```

where we have taken the initial speed and position to be some small numbers.[4] We still have no displayed results. We want to make a plot of displacement versus time. For this, we need a table of the variable and the function. The necessary commands are given below:

`Table[f[i],{i,m,n}]`	make a list of the values of f with i running from m to n
`Table[{x[i],y[i]},{i,m,n}]`	build an array of pairs $\{x_i, y_i\}$ with i ranging from m to n
`ListPlot[table]`	plot `table`, where `table` consists of pairs of numbers, the first one treated as the x-coordinate, the second one as y-coordinate

Recall that the last statement in `In[8]` creates the `x[i]`'s for the `t[i]`'s generated by `deltat[5,20]`. To put them in a table, we type in

```
In[9]:= tab=Table[{t[i],x[i]},{i,0,20}];
```

and tell *Mathematica* to plot the list:

the option **PlotJoined** for **ListPlot**

```
In[9]:= ListPlot[tab,PlotJoined->True];
```

The result—the usual parabola associated with a uniformly accelerated motion—is shown in Figure 3.3. The option `PlotJoined->True` draws a smooth curve through the points specified by the `Table`. Without this option, *Mathematica* simply plots the points.

The preceding routine works pretty well with many position-dependent forces. However, some of the most common forces such as the restoring force of a spring—Hooke's law—cause difficulties. This is due to the oscillatory nature of the solutions in such cases, and the inadequacy of Newton's method in finding the roots of oscillatory functions. This problem is only illusory, as the efficient way of treating motion is through differential equations, to which we shall return later.

3.2.2 Gravitational Fields

The gravitational field of a continuous mass distribution is given by

MM, p. 79

[4]If you take $v_0 = 0 = x_0$, *Mathematica* will give error and warning messages, because it will have difficulty evaluating the integrals.

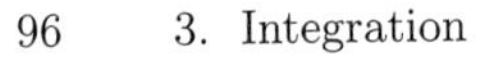

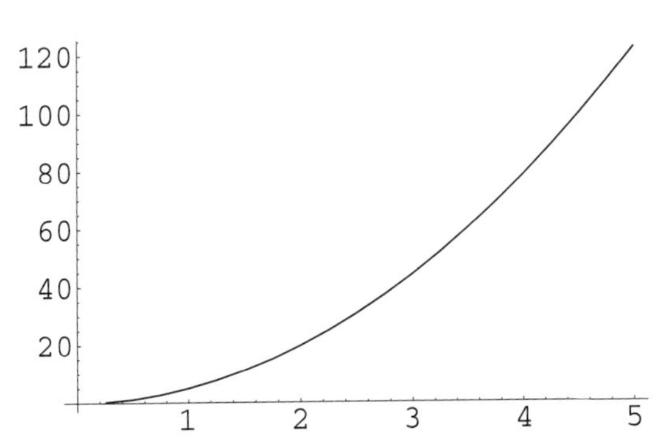

FIGURE 3.3. Uniformly accelerated motion as numerically calculated by *Mathematica* using Equation (3.2).

$$\mathbf{g}(\mathbf{r}) = -\iint_{\Omega} \frac{G\,dm(\mathbf{r}')}{|\mathbf{r}-\mathbf{r}'|^3}(\mathbf{r}-\mathbf{r}') \tag{3.3}$$

Field point is where the field is calculated; source point is where the element of mass is located.

where Ω is the region of integration, $\mathbf{r}$ is the position vector of the *field point*, $\mathbf{r}'$ is the position vector of the *source point*, and $\iint_{\Omega}$ is a generic notation for an integral that does not specify the dimensionality of integration—it could be a single, a double, or a triple integral.

Field of a Spheroid

The Earth is a sphere that is slightly flattened at the poles. For all practical purposes, we can ignore this flattening, because the difference between the radius at the equator and at the poles is extremely small, being 6378 km at the equator and 6357 at the poles. Nevertheless, it is a good numerical exercise to calculate the gravitational field of such a *spheroid*. Assuming that the shape of the Earth is described by

$$\frac{x^2}{a^2} + \frac{y^2}{a^2} + \frac{z^2}{b^2} = 1 \tag{3.4}$$

MM, pp. 125–126

we can express its gravitational field at a point (x, y, z) in terms of the following triple integral:

$$\mathbf{g}(\mathbf{r}) = \int_{-a}^{a} dx' \int_{-\sqrt{a^2-x'^2}}^{\sqrt{a^2-x'^2}} dy' \int_{-\frac{b}{a}\sqrt{a^2-x'^2-y'^2}}^{\frac{b}{a}\sqrt{a^2-x'^2-y'^2}} dz' \frac{\langle x-x', y-y', z-z'\rangle}{[(x-x')^2+(y-y')^2+(z-z')^2]^{3/2}} \tag{3.5}$$

From this integral we can calculate the three components of the gravitational field at any point (x, y, z). However, we are mostly interested in two specific points: $(a, 0, 0)$ on the equator, and $(0, 0, b)$ at the North Pole.

First let us concentrate on the equatorial point, for which there is only one component, g_x, which is given by

$$g_x = \int_{-a}^{a} dx' \int_{-\sqrt{a^2-x'^2}}^{\sqrt{a^2-x'^2}} dy' \int_{-\frac{b}{a}\sqrt{a^2-x'^2-y'^2}}^{\frac{b}{a}\sqrt{a^2-x'^2-y'^2}} dz' \frac{x-x'}{[(x-x')^2+(y-y')^2+(z-z')^2]^{3/2}} \tag{3.6}$$

The z' integration can be done analytically. So, rather than having *Mathematica* do the entire triple integral, we first ask it to do the indefinite integral and then substitute the limits. This is because, as mentioned earlier, *Mathematica* is more efficient in evaluating certain indefinite integrals than their definite counterparts. To start, we type in

important use of **Evaluate** in integration

```
In[1]:= f1[x_,xp_,y_,yp_,z_,zp_]=Evaluate[
        Integrate[((x-xp)^2+(y-yp)^2+(z-zp)^2)^3/2,zp]];
```

We have `Evaluate`d the integration so that we can substitute for z' (or zp) the upper and lower limits of integration. Without `Evaluate` *Mathematica* will substitute the new values of zp *before integration* and will get confused.

Now we evaluate the integrand of the remaining two variables. This involves evaluating $f1$ at the two limits of zp integration and multiplying the result by $x - x'$. So, we type in

```
In[2]:= fx[x_,xp_,y_,yp_,z_,a_,b_]=(x-xp)
        (f1[x,xp,y,yp,z,(b/a)Sqrt[a^2-xp^2-yp^2]]-
        f1[x,xp,y,yp,z,-(b/a)Sqrt[a^2-xp^2-yp^2]])
```

Finally, we integrate fx numerically to find g_x:

```
In[3]:= gx[x_,y_,z_,a_,b_]=NIntegrate[fx[x,xp,y,yp,z,a,b],
          {xp,-a,a},{yp,-Sqrt[a^2-xp^2],Sqrt[a^2-xp^2]}]
```

where we have ignored the constant $-G\rho$ in front of the integral.

At the North Pole, the surviving component is g_z. Thus, we have to evaluate

$$g_z = \iiint \frac{z-z'}{[(x-x')^2+(y-y')^2+(z-z')^2]^{3/2}} dx'dy'dz' \tag{3.7}$$

and the natural tendency is to perform the z' integration first. However, this approach will cause such severe singularities in the process of integration that *Mathematica* will quit the calculation. Switching the order of integration will alleviate the problem, and *Mathematica* will sail through the calculation with only minor inconvenience; another indication of the fact that integration is an art with lots of tricks to get acquainted with. So

another example of the subtleties of integration in *Mathematica* we perform the x' integration first and leave the z' integration for last and write

$$g_z = \int_{-b}^{b} (z - z')\, dz' \int_{-\sqrt{a^2-(az'/b)^2}}^{\sqrt{a^2-(az'/b)^2}} dy'$$
$$\int_{-\sqrt{a^2-y'^2-(az'/b)^2}}^{\sqrt{a^2-y'^2-(az'/b)^2}} \frac{dx'}{[(x-x')^2 + (y-y')^2 + (z-z')^2]^{3/2}} \tag{3.8}$$

Thus, as in the previous case, we type in

```
In[4]:= h1[x_,xp_,y_,yp_,z_,zp_]=Evaluate[
        Integrate[((x-xp)^2+(y-yp)^2+(z-zp)^2)^3/2,xp]];
```

Then evaluate the integrand of the remaining two variables. This involves evaluating $h1$ at the two limits of xp integration and multiplying the result by $z - z'$. So, we type in

```
In[5]:= fz[x_,y_,yp_,z_,zp_,a_,b_]=(x-xp)
        (h1[x,xp,y,yp,z,Sqrt[a^2-yp^2-(a zp/b)^2]]-
        h1[x,xp,y,yp,z,-Sqrt[a^2-yp^2-(a zp/b)^2]])
```

For details of finding correct limits of integration, see *MM*, pp. 93–96.

Finally, we integrate fz numerically to find g_z:

```
In[6]:= gz[x_,y_,z_,a_,b_]=NIntegrate[fx[x,y,yp,z,zp,a,b],
        {zp,-b,b},{yp,-Sqrt[a^2-(a zp/b)^2],
        Sqrt[a^2-(a zp/b)^2]}]
```

An interesting problem is to find the ratio a/b for which the field at the pole is maximum when the mass of the spheroid is held fixed and the density is uniform. Since the volume of the spheroid is $\frac{4}{3}\pi a^2 b$, for a spheroid of density ρ, we have

$$M = \frac{4}{3}\pi\rho a^2 b \quad \text{or} \quad a = \sqrt{\frac{3M}{4\pi\rho b}}$$

Ignoring the constants, we let $a = 1/\sqrt{b}$. At the pole, $x = 0 = y$ and $z = b$, but to avoid singularities, we let z be just a little larger than b, say $z = 1.0001b$. We can now plot g_z as a function of b. Here is the command that will accomplish that:

```
In[7]:= Plot[Chop[gz[0, 0, 1.0001 b, 1/Sqrt[b], b]],
        {b, 0.1, 1}]
```

Use **Chop** to get rid of the tiny imaginary part of a calculation result.

The command `Chop` gets rid of any negligible imaginary part that may accompany the result of integration. Figure 3.4 shows the plot of g_z at the North Pole as a function of the pole radius when the volume is fixed. It is interesting to note that there *is* indeed a b for which the gravitational field

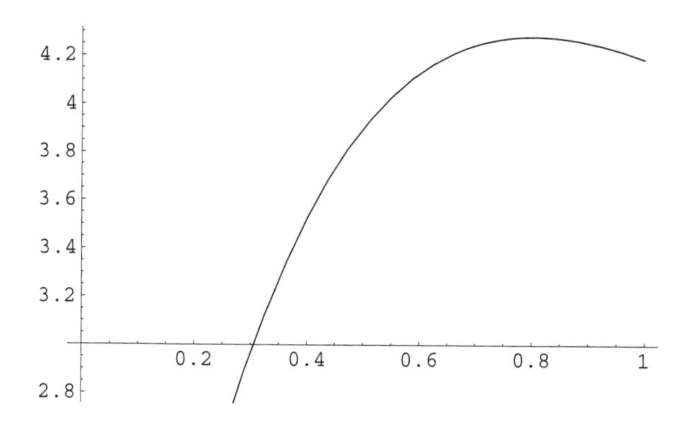

FIGURE 3.4. Gravitational acceleration (field) at the North Pole as a function of the pole radius b when the volume (or mass) is fixed.

is maximum and that the value of b is around 0.8. One may suspect that the value should be at $b = a = 1$, corresponding to a sphere,[5] but one has to realize that a large value of b corresponds to a larger distance from the center of mass and, therefore, a smaller field.

For Earth, $a = 6378$ km and $b = 6357$ km, so that the ratio a/b becomes 1.0033. With $b = 1$, we type in

```
In[8]:= gx[1.003301, 0, 0, 1.0033, 1]
```

and obtain $4.19706 + 2.87295 \times 10^{-30}$ I. The small imaginary part is what `Chop` gets rid of. Note that we have increased x slightly above a to avoid singularities. Similarly, typing in

```
In[9]:= gz[0, 0, 1.000001, 1.0033, 1]
```

will produce $4.19983 + 2.92009 \times 10^{-30}$ I. These are not actual values for accelerations. To obtain the two values of g at the North Pole and at the equator, we need to insert back the constants we ignored earlier. Since we have set $b = 1$, all lengths are measured in terms of b. In particular, g_x and g_z, which have dimension of length, are also measured in units of b. The result of the integration in the two cases can be written as bA, where A is one of the two numbers obtained above. We note that

$$g = G\rho bA = G\frac{M}{\frac{4}{3}\pi a^2 b} bA = \frac{GM}{a^2}\frac{3A}{4\pi}$$

[5] $b = a$ and $a = 1/\sqrt{b}$ imply that $a = b = 1$.

With $G = 6.67\times10^{-11}$ and $M = 5.98\times10^{24}$, we can calculate the equatorial and polar accelerations:

$$g_{\text{pole}} = \frac{(6.67\times10^{-11})(5.98\times10^{24})}{(6.378\times10^{6})^2} \times \frac{3\times 4.19983}{4\pi} = 9.8311 \text{ m/s}^2$$

$$g_{\text{equator}} = \frac{(6.67\times10^{-11})(5.98\times10^{24})}{(6.378\times10^{6})^2} \times \frac{3\times 4.19706}{4\pi} = 9.8246 \text{ m/s}^2$$

As expected, the polar acceleration is slightly larger than the equatorial acceleration because poles are at lower "altitudes" than the equator. The actual accelerations differ slightly from the two numbers given above, partly due to the variation in density both locally and throughout the interior of the Earth.

Evidence for Dark Matter

Gravity is the dominant force holding the large-scale structures in the universe, from planets to galaxies. This force acts between all objects that have mass (or energy, using the equivalence of the two in relativity theory) and causes them to accelerate. Since many celestial accelerations in the universe are of centripetal type, a measurement of speed can give us an indication of the nature of matter causing the acceleration.

Here we want to examine the acceleration of stars in a galaxy such as the Milky Way assuming it has a very simple disklike structure with negligible thickness and uniform surface mass density σ. The appropriate coordinates to use for integration are cylindrical. Using Equation (3.3) and the following information

MM, p. 18; p. 112 with $z = 0 = z'$; figure on p. 113, with $\varphi = 0$

$$\mathbf{r} = \rho\hat{\mathbf{e}}_\rho, \quad \mathbf{r}' = \rho'\hat{\mathbf{e}}_{\rho'}, \quad \hat{\mathbf{e}}_\rho\cdot\hat{\mathbf{e}}_{\rho'} = \cos\varphi'$$

in which we assume that our field point is on the x-axis, one can show that

$$g_x = -G\sigma\int_0^{2\pi} d\varphi' \int_0^a d\rho' \frac{\rho'(\rho-\rho'\cos\varphi')}{(\rho^2+\rho'^2-2\rho\rho'\cos\varphi')^{3/2}}$$

$$g_y = 0 = g_z$$

Thus, the nonvanishing component of the field becomes

$$g_x = -G\sigma\rho\int_0^{2\pi} d\varphi' \underbrace{\int_0^a \frac{\rho'\,d\rho'}{(\rho^2+\rho'^2-2\rho\rho'\cos\varphi')^{3/2}}}_{\equiv int1}$$
$$+\, G\sigma\int_0^{2\pi} d\varphi'\cos\varphi' \underbrace{\int_0^a \frac{\rho'^2\,d\rho'}{(\rho^2+\rho'^2-2\rho\rho'\cos\varphi')^{3/2}}}_{\equiv int2} \tag{3.9}$$

We see that the problem boils down to the calculation of two double integrals.

We calculate the inner integral of each one and add the results. For the first inner integral we type in

```
In[1]:= int1[r_,phi_,a_]:=Evaluate[Integrate[t/(t^2-
        2 r t Cos[phi]+r^2)^3/2],{t,0,a}]
```

where we have used `r` for ρ, `phi` for φ', and `t` for ρ'. *Mathematica* calculates the integral fairly quickly. However, if we do the same thing for the second integral, *Mathematica* gets stuck. So we try evaluating the indefinite integral by typing in

```
In[2]:= f2[r_,phi_,t_]:=Evaluate[Integrate[t^2/(t^2-
        2 r t Cos[phi]+r^2)^3/2],t];
        int2[r_,phi_,a_]:=f2[r,phi,a]-f2[r,phi,0]
```

Now we add these two integrals—after multiplying by appropriate factors and ignoring the factor $-G\sigma$ in front of the double integrals—to obtain the integrand of the φ' integration:

```
In[3]:= h[r_,phi_,a_]:=Simplify[r int1[r,phi,a]
        -Cos[phi]int2[r,phi,a]]
```

Then, typing `h[r,phi,a]` produces

$$\frac{\sqrt{r^2}}{r} - \frac{r}{\sqrt{a^2+r^2-2ar\cos\varphi}}$$
$$+ \cos\varphi\Big(\frac{2a}{\sqrt{a^2+r^2-2ar\cos\varphi}} + \mathrm{Log}\left[\sqrt{r^2} - r\cos\varphi\right]$$
$$- \mathrm{Log}\left[a - r\cos\varphi + \sqrt{a^2+r^2-2ar\cos\varphi}\right]\Big)$$

which we simplify manually and write as

$$h(r,\varphi,a) = 1 - \frac{r-2a\cos\varphi}{\sqrt{a^2+r^2-2ar\cos\varphi}} + \cos\varphi\,\mathrm{Log}\left[r - r\cos\varphi\right]$$
$$- \cos\varphi\,\mathrm{Log}\left[a - r\cos\varphi + \sqrt{a^2+r^2-2ar\cos\varphi}\right] \qquad (3.10)$$

What is left now is to integrate $h(r,\varphi,a)$ over φ. We do this term by term, and try to integrate each term *analytically*, using as many tricks as possible to give results. The first term is trivial and gives $gx1 = 2\pi$. For the second term, we type in

```
In[4]:= gx2[r_,a_]:=Simplify[Integrate[
        (r-2 a Cos[phi])/Sqrt[t^2- 2 r t Cos[phi]+r^2],
        {phi,0,2 Pi}]]
```

to which *Mathematica* responds by giving

$$\frac{4\left((a-r)^2 \operatorname{EllipticE}\left[-\frac{4ar}{(a-r)^2}\right] - a^2 \operatorname{EllipticK}\left[-\frac{4ar}{(a-r)^2}\right]\right)}{\sqrt{(a-r)^2 r}}$$

The fact that the output is in terms of the so-called *elliptic functions* should be of no concern to us.

MM, Section 6.1.4 discusses elliptic functions briefly.

The terms involving (natural) logarithm are more difficult to handle. A direct request to *Mathematica* will yield no result. Even indefinite integration will produce no response. However, if we integrate by parts—and in the process get rid of the log terms—we may have some luck. Recall that in integration by parts, one integrates the differential identity $d(uv) = u\,dv + v\,du$ to obtain $\int_a^b u\,dv = uv|_a^b - \int_a^b v\,du$. A judicious choice of u and v may simplify the integration on the right-hand side considerably. With this in mind, we type in

```
In[5]:= u1[r_,phi_,a_]:=Log[r-r Cos[phi]];
        v[r_,phi_,a_]:=Evaluate[Integrate[Cos[phi],phi]]
```

We can evaluate $uv|_a^b$ as follows:

```
In[6]:= uv1[r_,a_]:= Limit[u1[r,phi,a]v[r,phi,a],
        phi->2 Pi]-Limit[u1[r,phi,a]v[r,phi,a],phi->0]
```

The reason for the appearance of `Limit` is that the direct substitution will give an indeterminate result. Typing `uv1[r,a]` gives 0, showing that the $uv|_a^b$ part does not contribute to the integral. The remaining part, which we call `intuv1`, is calculated by typing in

```
In[7]:= intuv1[r_,phi_,a_]:=
        -Simplify[Evaluate[D[u1[r,phi,a],phi]] v[r,phi,a]]
```

and integrating the result

```
In[8]:= gx3=Integrate[intuv1[r, phi,a],phi,0,2 Pi]
```

yielding -2π as the final answer.

Now we proceed with the second log term. Since v is already defined, we need only to define u, which we name $u2$:

```
In[9]:= u2[r_,phi_,a_]:=
        Log[a - r Cos[phi] + Sqrt[a^2+r^2-2 a r Cos[phi]]
```

For the term $uv|_a^b$, we do as before and obtain zero. The remaining part, which we call `intuv2`, is

```
In[10]:= intuv2[r_,phi_,a_]=
         Simplify[Evaluate[D[u2[r, phi,a],phi]] v[r,phi,a]]
```

whose output is

$$-\frac{r\left(1+\frac{a}{\sqrt{a^2+r^2-2ar\cos\varphi}}\right)\sin^2\varphi}{a-r\cos\varphi+\sqrt{a^2+r^2-2ar\cos\varphi}}$$

or

$$-\frac{r\sin^2\varphi}{a-r\cos\varphi+\sqrt{a^2+r^2-2ar\cos\varphi}}-\frac{\frac{ra\sin^2\varphi}{\sqrt{a^2+r^2-2ar\cos\varphi}}}{a-r\cos\varphi+\sqrt{a^2+r^2-2ar\cos\varphi}}$$

We integrate these separately by typing in

```
In[11]:= gx4[r_,a_]:=Simplify[-r Integrate[Sin[phi]^2
         /(a-r Cos[phi]+Sqrt[a^2+r^2-2a r Cos[phi])]];
         gx5[r_,a_]:=Simplify[-r a Integrate[Sin[phi]^2
         Sqrt[a^2+r^2-2a r Cos[phi]])]/(a-r Cos[phi]+
         Sqrt[a^2+r^2-2a r Cos[phi]])]]
```

To find the gravitational field, we add $gx1$ through $gx5$:

```
In[12]:= gx[r_,a_]:=Simplify[gx1+gx2[r,a]+gx3+
         gx4[r,a]+gx5[r,a]]
```

with the final result [after changing back to the notation used in Equation (3.9)]

$$\begin{aligned} g_x(\rho, a) = &\frac{2(a-\rho)^2}{\rho\sqrt{(a-\rho)^2}} \text{EllipticE}\left[-\frac{4a\rho}{(a-\rho)^2}\right] \\ &-\frac{2(a^2+\rho^2)}{\rho\sqrt{(a-\rho)^2}} \text{EllipticK}\left[-\frac{4a\rho}{(a-\rho)^2}\right] \end{aligned} \tag{3.11}$$

Now that we have the field as a function of distance from the center, we can plot it. The command `Plot[g[r,1],{r,0,2}]` will produce the plot of Figure 3.5. The field starts at zero at the center, as expected; increases when approaching the rim, as expected; and falls off outside when moving away from the disk, as expected. What seems to be surprising is the fact that the field goes to infinity at the rim. Although this phenomenon is not relevant to our present discussion, a detour into explaining it is worthwhile.

It is well known that the gravitational (or electrostatic, as the two obey the same mathematical laws) field of a hollow spherical shell is zero inside if the mass (or charge) is distributed uniformly on the shell. Although this can be proved rigorously, a simple intuitive argument will shed light on the infinity encountered in the case of a disk. Let point P be anywhere inside a spherical shell as shown in Figure 3.6. Draw a diameter passing through P (the dashed line in the figure), and construct the two cones—one larger,

MM, pp. 121–122 prove that the field inside a hollow spherical shell is zero.

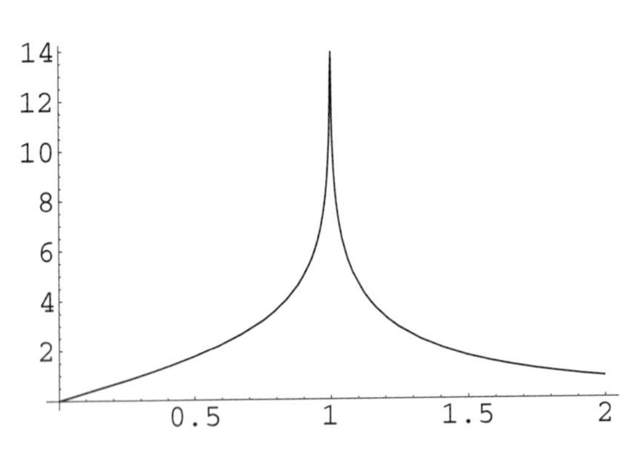

FIGURE 3.5. Gravitational acceleration (field) at a distance ρ from the center of a disk. The disk is assumed to have a unit radius.

one smaller, but both *infinitesimal*—with the diameter as their common axis. Let h_1 denote the height of the small cone and h_2 that of the large one. Then the field at P due to AC is

$$\frac{G\Delta m_1}{h_1^2} = \frac{G\sigma\Delta a_1}{h_1^2} = G\sigma\Delta\Omega_1$$

where σ is the surface mass density and $\Delta\Omega_1$ is the solid angle subtended by the cone at P. A similar calculation shows that the field due to BD is $G\sigma\Delta\Omega_2$, where $\Delta\Omega_2$ is the solid angle subtended by the larger cone at P. But the two solid angles are equal; so the two fields are equal, but in the opposite directions. Therefore, inside, the field will be zero.

We can even calculate the field right at the surface outside. To do so, let's move P infinitesimally close to the surface. Then, the argument above still holds, as long as P is inside. Just outside, the fields are still equal, but *in the same direction.* So,

$$g_{\text{out}} = 2G\sigma\Delta\Omega_1 = 2G\sigma(2\pi) = 4\pi G\sigma$$

because the solid angle subtended by AC when P is infinitesimally close to it is 2π.[6] We can rewrite the last equation as

$$g_{\text{out}} = 4\pi G\frac{M}{4\pi a^2} = \frac{GM}{a^2}$$

where M is the total mass of the shell and a its radius. This is the familiar result that (just) outside a uniform spherical shell, the gravitational field is the same as that of a point (with equal) mass located at the center of the shell.

[6]This is analogous to the fact that the angle subtended by a line segment at a point P approaches π as P approaches the line segment.

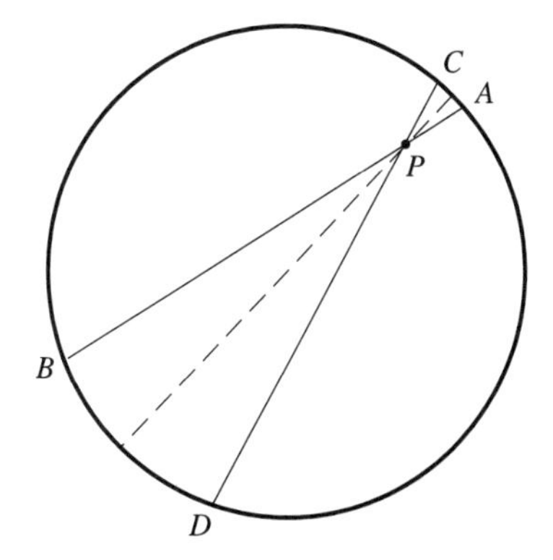

FIGURE 3.6. For a spherical shell, the contributions from AC and BD cancel at the boundary inside, but add at the boundary outside. For a ring, the contributions from AC is larger than that from BD, and gets larger and larger as P gets closer and closer to AC.

Now let us study the behavior of the field inside and outside a *ring*. Again refer to Figure 3.6, but assume that the circle is a ring rather than a spherical shell. Then as before, we can write the contribution of the two arcs of the circle to the gravitational field:

$$g_1 = \frac{G\Delta m_1}{h_1^2} = \frac{G\lambda\Delta l_1}{h_1^2} = \frac{G\lambda}{h_1}\frac{\Delta l_1}{h_1} = \frac{G\lambda\Delta\theta}{h_1}$$
$$g_2 = \frac{G\Delta m_2}{h_2^2} = \frac{G\lambda\Delta l_2}{h_2^2} = \frac{G\lambda}{h_2}\frac{\Delta l_2}{h_2} = \frac{G\lambda\Delta\theta}{h_2}$$

where λ is the linear mass density and $\Delta\theta$ is the common angle subtended by AC and BD at P. This result shows clearly that as P gets closer and closer to the ring, g_1 gets larger and larger without bound, while g_2 approaches the finite value of $G\lambda\pi/(2a)$.

Thus, we should not be surprised if the gravitational field blows up at the rim of an infinitely thin disk. A real disk, of course, has a thickness and no sharp edges, and its gravitational field will be finite and well-defined at all points in space.

Since we are interested only in the field outside, we need not worry about the infinity at the rim. If there is an object orbiting the disk at the distance ρ, it will have a speed given by $g = v^2/\rho$. With g given by Equation (3.11), we find the speed as

$$v^2 = \frac{-2G\sigma}{\sqrt{(a-\rho)^2}}\Big\{(a-\rho)^2 \,\text{EllipticE}\left[-\frac{4a\rho}{(a-\rho)^2}\right] - (a^2+\rho^2)\,\text{EllipticK}\left[-\frac{4a\rho}{(a-\rho)^2}\right]\Big\} \tag{3.12}$$

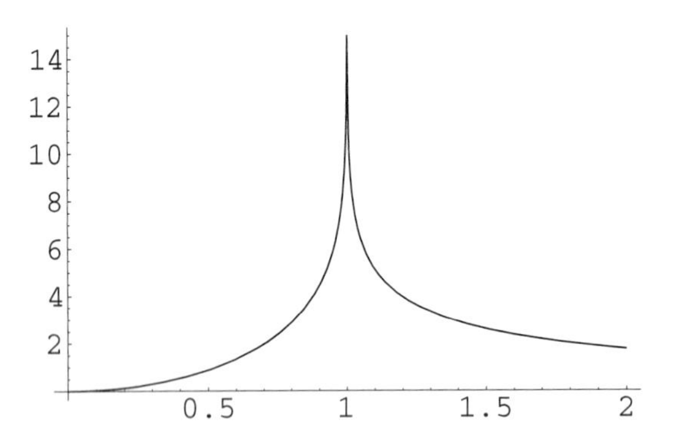

FIGURE 3.7. The speed decreases as the distance from the center of galaxy increases.

where we have restored the constant factor we ignored at the beginning of our calculation. Plotting v as a function of ρ produces the graph of Figure 3.7, which shows a drop in v with distance from the center of the disk.

evidence for dark matter

Observation of many galaxies has indicated a fairly constant v for values of ρ in the range of tens of millions of light years, way beyond the visible edge of the galaxy, typically in the range of a hundred thousand light years. In fact, if there is any change in speed with distance, it is in the form of an *increase*! The only way this can happen—as an examination of Equation (3.12) reveals instantly—is for σ not to stop at the visible edge of the galaxy, but to continue its presence at larger distances, and this conclusion does not depend on the shape of the galaxy. A spherical galaxy, or that of any other shape, will lead to the same conclusion. However, this presence cannot be visible, because the *visible* size of the galaxy is set by optical observations. The only solution to this riddle is the existence of **dark matter**, a form of matter that emits no electromagnetic signal but makes its presence felt by its gravitational attraction.

Measurement of speed of objects attracted by nearby galaxies is only one way of inferring the existence of dark matter. Other cosmological indicators signal the presence of dark matter all over the universe. There is no doubt that dark matter exists and that it constitutes the majority—about 90%—of the mass of the universe.

3.3 Integration in Electrostatics

MM, pp. 79 and 377

The electric field and electrostatic potential of a continuous charge distribution are given by

$$\mathbf{E}(\mathbf{r}) = -\iint_{\Omega} \frac{k_e\, dq(\mathbf{r}')}{|\mathbf{r}-\mathbf{r}'|^3}(\mathbf{r}-\mathbf{r}') \tag{3.13}$$

and

$$\Phi(\mathbf{r}) = -\iint_{\Omega} \frac{k_e\, dq(\mathbf{r}')}{|\mathbf{r}-\mathbf{r}'|} \tag{3.14}$$

respectively, where Ω, $\mathbf{r}$, and $\mathbf{r}'$ are as in Equation (3.3).

Let us consider a linear charge distribution (a curve), described—in Cartesian coordinates—parametrically by

$$\mathbf{r}' \equiv \langle x', y', z'\rangle = \langle f(t), g(t), h(t)\rangle, \qquad a \le t \le b$$

where $t = a$ gives the initial point of the curve and $t = b$ its final point. If the linear charge density is given by a function $\lambda(x', y', z')$, then

$$dq(\mathbf{r}') = \lambda(x', y', z')\sqrt{(dx')^2 + (dy')^2 + (dy')^2}$$

or

$$dq(\mathbf{r}') = \underbrace{\lambda(f(t), g(t), h(t))}_{\text{Call this } \Lambda(t)} \sqrt{[f'(t)]^2 + [g'(t)]^2 + [h'(t)]^2}\, dt$$

where f', g', and h' denote derivatives.

When we substitute all the above in Equation (3.13) and separate the components of the field, we get

See *MM*, p. 104.

$$\begin{aligned}
E_x &= \int_a^b \frac{k_e\Lambda(t)\sqrt{[f'(t)]^2 + [g'(t)]^2 + [h'(t)]^2}\,[x - f(t)]}{\left\{[x - f(t)]^2 + [y - g(t)]^2 + [z - h(t)]^2\right\}^{3/2}}\, dt \\
E_y &= \int_a^b \frac{k_e\Lambda(t)\sqrt{[f'(t)]^2 + [g'(t)]^2 + [h'(t)]^2}\,[y - g(t)]}{\left\{[x - f(t)]^2 + [y - g(t)]^2 + [z - h(t)]^2\right\}^{3/2}}\, dt \\
E_z &= \int_a^b \frac{k_e\Lambda(t)\sqrt{[f'(t)]^2 + [g'(t)]^2 + [h'(t)]^2}\,[z - h(t)]}{\left\{[x - f(t)]^2 + [y - g(t)]^2 + [z - h(t)]^2\right\}^{3/2}}\, dt
\end{aligned} \tag{3.15}$$

Similarly, substituting the same information in Equation (3.14) yields

See *MM*, p. 105.

$$\Phi(x, y, z) = \int_a^b \frac{k_e\Lambda(t)\sqrt{[f'(t)]^2 + [g'(t)]^2 + [h'(t)]^2}}{\left\{[x - f(t)]^2 + [y - g(t)]^2 + [z - h(t)]^2\right\}^{1/2}}\, dt \tag{3.16}$$

In particular, if the density is uniform, and we are interested in the potential in the xy-plane of a curve that also lies in the xy-plane, then $z = 0$ and $h(t) = 0$, and we have

$$\Phi(x, y) = k_e\lambda \int_a^b \frac{\sqrt{[f'(t)]^2 + [g'(t)]^2}}{\left\{[x - f(t)]^2 + [y - g(t)]^2\right\}^{1/2}}\, dt \tag{3.17}$$

where λ is the constant linear density.

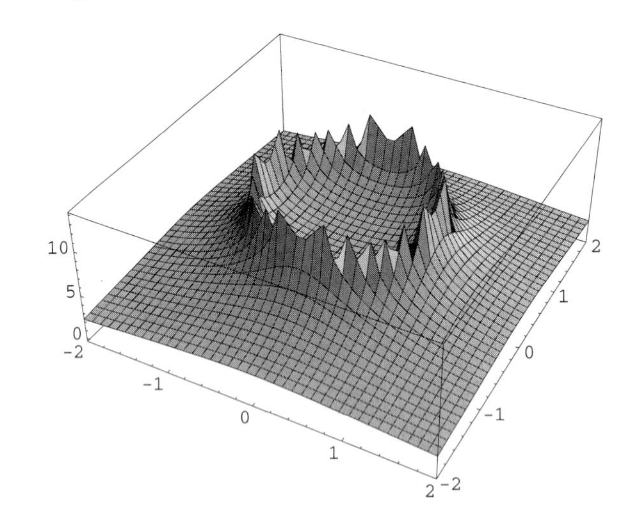

FIGURE 3.8. The potential of a circular loop of uniformly charged ring in the xy-plane. Note that at the ring the potential is infinite.

3.3.1 Potential of a Ring

Now consider a uniformly charged circular ring of unit radius lying in the xy-plane. For such a charge distribution

$$f(t) = \cos t, \quad g(t) = \sin t, \quad 0 \le t \le 2\pi$$

Thus, Equation (3.17) becomes

$$\Phi(x, y) = \int_0^{2\pi} \frac{1}{\left\{[x - \cos t]^2 + [y - \sin t]^2\right\}^{1/2}} \, dt \tag{3.18}$$

in which we have ignored the constants outside the integral. We want to plot this potential as a function of x and y. The *Mathematica* command for the *numerical* integration of the potential is

```
In[1]:= f[x_,y_]:=NIntegrate[1/Sqrt[(x-Cos[t])^2
          +(y-Sin[t])^2],{t,0,2 Pi}],
```

and its three-dimensional plot can be obtained using

```
In[2]:= Plot3D[f[x,y],{x,-2,2},{y,-2,2}, PlotPoints->40]
```

The option `PlotPoints->40` increases the smoothness of the plot. Figure 3.8 shows such a plot, where the boundary of the ring—at which the potential is infinite—is markedly conspicuous.

3.3.2 Potential of a Spiral

Another interesting example is the potential of a uniformly charged flat spiral in the xy-plane. The equation of a spiral is normally given in polar coordinates. Suppose the polar equation of this spiral is $r = 0.5\theta$. The Cartesian equations of the spiral is then [see also Equation (1.3)]

$$x(\theta) = r\cos\theta = 0.5\theta\cos\theta \Rightarrow f(t) = 0.5t\cos t$$
$$y(\theta) = r\sin\theta = 0.5\theta\sin\theta \Rightarrow g(t) = 0.5t\sin t$$

Let us calculate the potential at the points of a plane parallel to the xy-plane crossing the z-axis at $z = 0.5$. Evaluating f' and g', substituting in Equation (3.16), and simplifying—using *Mathematica* or otherwise—yields

$$\Phi(x, y, 0.5) = k_e\lambda \int_0^{20} \frac{0.5\sqrt{1+t^2}}{\left\{[x - 0.5t\cos t]^2 + [y - 0.5t\sin t]^2 + [0.5]^2\right\}^{1/2}}\, dt \tag{3.19}$$

where we have taken the final point of the parameter t to be 20 so that the spiral can turn a few times. Ignoring the constants in front of the integral, the *Mathematica* inputs for plotting the potential are now

```
In[3]:= f[x_,y_]:=NIntegrate[Sqrt[1+t^2]/Sqrt[(x-0.5 t
          Cos[t])^2+(y-0.5 t Sin[t])^2+0.25],{t,0,20}],
```

and

```
In[4]:= Plot3D[f[x,y],{x,-8,8},{y,-8,8}, PlotPoints->40]
```

Figure 3.9 shows the three-dimensional plot of the potential as a function of x and y. Notice that the potential is not infinite anywhere, because the points of the new plane never touch the spiral.

3.3.3 Flat Surface Charge Distributions

If the electric charge is distributed in a region Ω of the xy-plane with a known surface charge density σ, we can calculate its electric field and potential at an arbitrary point in space. In fact, it is easy to show that

MM, p. 116

$$\begin{aligned} E_x &= k_e \iint_\Omega \frac{\sigma(x', y')(x - x')\, dx'\, dy'}{\{(x - x')^2 + (y - y')^2 + z^2\}^{3/2}} \\ E_y &= k_e \iint_\Omega \frac{\sigma(x', y')(y - y')\, dx'\, dy'}{\{(x - x')^2 + (y - y')^2 + z^2\}^{3/2}} \\ E_z &= k_e z \iint_\Omega \frac{\sigma(x', y') dx'\, dy'}{\{(x - x')^2 + (y - y')^2 + z^2\}^{3/2}} \end{aligned} \tag{3.20}$$

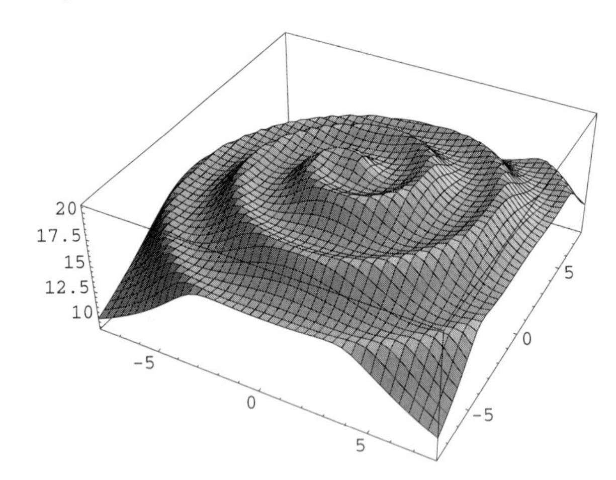

FIGURE 3.9. The potential of a uniformly charged flat spiral in a plane parallel to the xy-plane. Note the outline of the spiral at which the potential is maximum.

and

$$\Phi(x,y,z) = k_e \iint_\Omega \frac{\sigma(x',y')dx'\,dy'}{\sqrt{(x-x')^2+(y-y')^2+z^2}} \tag{3.21}$$

where x' and y' run over the points of Ω in the double integrals. For example, for a uniformly charged square of side $2a$ and surface charge density σ, whose center is the origin, the potential is

$$\begin{aligned} \Phi(x,y,z) &= k_e\sigma \int_{-a}^{a}\int_{-a}^{a} \frac{dx'\,dy'}{\sqrt{(x-x')^2+(y-y')^2+z^2}} \\ &= k_e\sigma \int_{-a}^{a} \ln\left[\frac{a+y+\sqrt{(x-x')^2+(y+a)^2+z^2}}{-a+y+\sqrt{(x-x')^2+(y-a)^2+z^2}}\right] dx' \end{aligned} \tag{3.22}$$

where we have observed the recommended practice of carrying out one of the integrations analytically, leaving only a single integral for numerical calculation.

3.4 Integration in Magnetism

MM, p. 111

The calculation of the magnetic fields of current carrying wires is done using the Biot–Savart law. This law is necessarily written in terms of an integral, because unlike the electric and gravitational fields, whose mathematical expressions involve local point charges and masses, magnetic fields are caused by *extended* electric currents, and, as such, *require* integrals.

Biot–Savart law for a current-carrying wire

The Biot–Savart law for a filament carrying a current I at a point P

with position vector $\mathbf{r}$ is given by

$$\mathbf{B}(\mathbf{r}) = k_m I \int \frac{d\,\mathbf{r}' \times (\mathbf{r} - \mathbf{r}')}{|\mathbf{r} - \mathbf{r}'|^{3/2}} \tag{3.23}$$

where the integral is over the coordinates of points of the filament. In the numerical calculations below, we ignore the multiplicative constant $k_m I$.

3.4.1 Circular Loop

Let us start with the simple example of a current-carrying circular loop of radius a located in the xy-plane. With the center of the circle at the origin of a cylindrical coordinate system, the three cylindrical components of the magnetic field are

MM, pp. 113–114

$$\begin{aligned} B_\rho &= k_m I a z \int_0^{2\pi} \frac{\cos t\, dt}{(\rho^2 + a^2 - 2a\rho\cos t + z^2)^{3/2}} \\ B_\varphi &= 0 \\ B_z &= -k_m I a \int_0^{2\pi} \frac{(\rho\cos t - a)\, dt}{(\rho^2 + a^2 - 2a\rho\cos t + z^2)^{3/2}} \end{aligned} \tag{3.24}$$

We type the nonvanishing components in *Mathematica*:

```
In[1]:= Br[r_,z_]:=z Integrate[Cos[t]/(r^2+1-2rCos[t]+
        z^2)^3/2,{t,0,2Pi}]
        Bz[r_,z_]:= Integrate[(rCos[t]-1)/(r^2+1-2rCos[t]+
        z^2)^3/2,{t,0,2Pi}]
```

where we have taken the radius of the loop to be unity and ignored the constants in front of the integral. We could have used `NIntegrate` instead; however, as it turns out, *Mathematica* can do the integrals analytically—in terms of the elliptic functions.

It is instructive to render a graph of the field lines of the magnetic field. To do so, we first have to load the special graphics package that draws field lines. This is done by first typing in `<<Graphics‘PlotField‘` and then the command

```
In[2]:= PlotVectorField[{Br[r,z],Bz[r,z]},{r,-2,2},
        {z,-1,1}]
```

The result is the diagram shown in Figure 3.10. The figure shows only a cross section of the field lines cut by a plane perpendicular to the loop and passing through its center. The cross section of the loop is shown as two empty spots halfway to the right and left of the middle vertical.

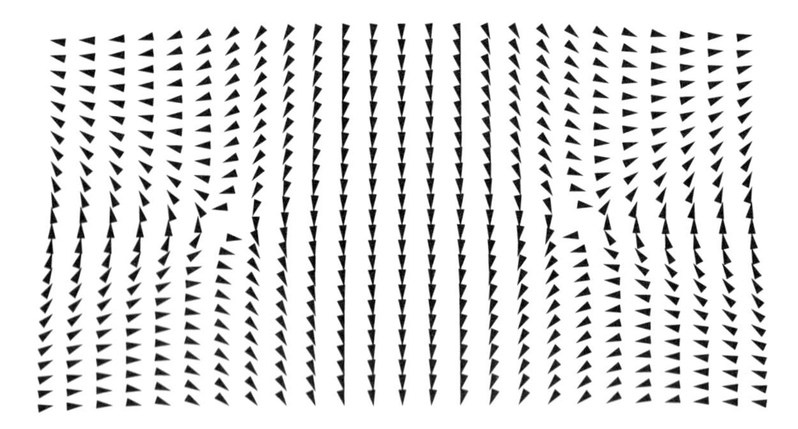

FIGURE 3.10. The magnetic field lines of a circular loop. The diagram shows only the lines as they appear on a plane perpendicular to the loop and passing through its center.

3.4.2 Current with General Shape

The example of a circular loop was very simple. We had the integrals at our disposal, and all we had to do was evaluate them. *Mathematica* can do much better than that. But it requires that we derive a general formula for the magnetic field of a current-carrying filament whose shape is quite arbitrary. We use Cartesian coordinates for simplicity.

Suppose that the filament is described in Cartesian coordinates by a parametric equation of the form

$$\mathbf{r}' = \langle x', y', z' \rangle = \langle f(t), g(t), h(t) \rangle$$

In *Mathematica* language this—and the coordinates of P—can be typed in as

```
In[1]:= rp[t_]:={f[t],g[t],h[t]}; r[x_,y_,z_]:={x,y,z}
```

The integrand of Equation (3.23), which is a vector, is typed in simply as

```
In[2]:= intB[x_,y_,z_,t_]:= Cross[rp'[t],r[x,y,z]-rp[t]]
        /((r[x,y,z]-rp[t]).(r[x,y,z]-rp[t]))^3/2
```

where instead of $d\mathbf{r}'$ we have used its derivative with respect to t, the latter being the variable of integration.

We are interested in the components of the magnetic field in Cartesian coordinates. To find these components, we first define our unit vectors:

```
In[3]:= ex={1,0,0}; ey={0,1,0}; ez={0,0,1};
```

MM, p. 6, Box 1.1.2

Then take the dot product of these vectors with `intB` to find the components of the integrand. So, we define three new integrands each corresponding to one component of $\mathbf{B}$:

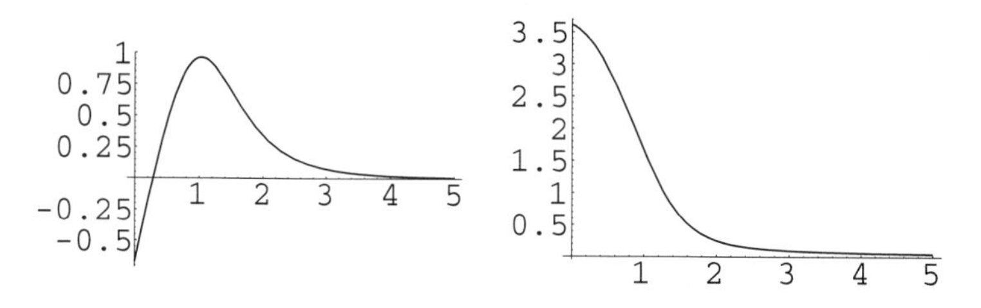

FIGURE 3.11. The x- (left) and z- (right) component of the magnetic field of a short solenoid of radius 1.

```
In[4]:= intBx[x_,y_,z_,t_]:=intB[x,y, t].ex;
        intBy[x_,y_,z_,t_]:=intB[x,y, t].ey;
        intBz[x_,y_,z_,t_]:=intB[x,y, t].ez;
```

and numerically integrate them to find the three components of the magnetic field:

```
In[5]:= Bx[x_,y_,z_,a_,b_]:=NIntegrate[intBx[x,y,z,t],
        {t,a,b}];
          By[x_,y_,z_,a_,b_]:=NIntegrate[intBy[x,y,z,t],
        {t,a,b}];
          Bz[x_,y_,z_,a_,b_]:=NIntegrate[intBz[x,y,z,t],
        {t,a,b}];
```

3.4.3 Solenoid

With the general formulas for the three components of **B** at our disposal, we can calculate the fields of currents of specific shape. One shape of interest is the helix, corresponding to a solenoid. Let us type in

```
In[6]:= f[t_]:=Cos[t]; g[t_]:=Sin[t]; h[t_]:=0.05 t;
        a=-5; b=5;
```

This describes a solenoid of radius 1, the beginning and end of whose filament occur at $(\cos(-5), \sin(-5), 0.05 \times (-5))$ [or $(0.284, -0.96, -0.25)$] and $(0.284, -0.96, 0.25)$, respectively. So, the length of the solenoid is 0.5, and the spacing between consecutive windings is $0.05 \times 2\pi = 0.314$. Hence, there are somewhat less than two windings on the solenoid.

Let us plot the x- and z-components of the magnetic field outside the solenoid. We choose a plane parallel to the xy-plane intersecting the z-axis at $z = 1$, and plot B_x and B_z as a function of x, i.e., for field points along the x-axis. Since we have a general expression for the field, and the functions describing the solenoid are all defined above, all we need to do is plot the functions.

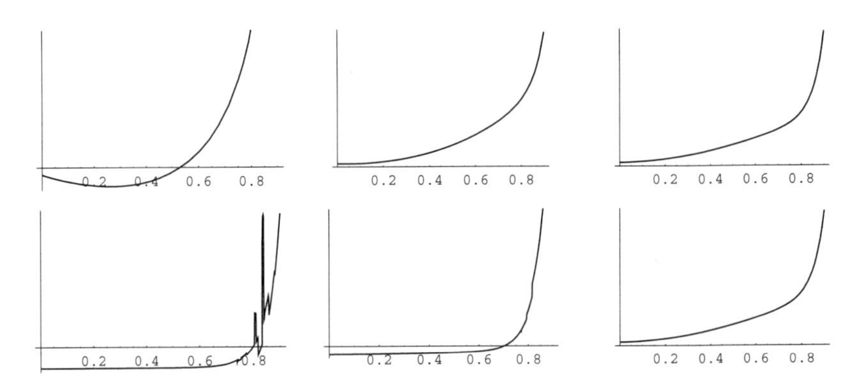

FIGURE 3.12. The z-component of the magnetic field inside a solenoid of radius 1. The upper-left diagram is the field of the solenoid when its length L is 0.5. Moving clockwise, we see B_z for $L = 1$, $L = 2$, $L = 5$, $L = 8$, and $L = 10$, respectively.

```
In[7]:= Plot[Bx[x,0,1,a,b],{x,0,5},PlotRange->All]
```

produces the plot on the left of Figure 3.11 and

```
In[8]:= Plot[Bz[x,0,1,a,b],{x,0,5},PlotRange->All]
```

yields the plot on the right. These plots demonstrate the behavior of components expected on physical grounds. For example, B_x is expected to be maximum when the point is right on top of the wire (almost a single loop) and B_z to be maximum at the center.

Now let us focus on points close to the xy-plane. In fact, let us find the components of the field *on* that plane. This will help us investigate the long-solenoid limit of the field, which is calculated in elementary physics courses using Ampere's circuital law. The reader recalls that for such a solenoid, the field inside is *constant* and entirely in the z-direction, and outside it is zero. It is the constancy of the field that we would like to investigate as the length of the solenoid increases.

We start with the short solenoid of length $L = 0.5$ as above. An attempt at calculating B_x or B_y will produce a complaint by *Mathematica* about the integral being oscillatory and converging too slowly, but the answer given out is almost zero. This "constant" magnetic field may be surprising, as the length of the solenoid is a only fraction of its radius, and hardly could be called "long." However, a very short solenoid can be approximated by a circular loop; and for a circular loop in the xy-plane, the field is entirely in the z-direction [see Equation (3.24)].

The z-component of the magnetic field can, however, be calculated numerically and plotted. The command

```
In[9]:= Plot[Bz[x,0,0,a,b],{x,0,0.9}]
```

produces the diagram shown in the upper-left corner of Figure 3.12. This figure clearly shows that B_z has some noticeable variation as one moves away from the axis of the solenoid towards its lateral surface. The rest of Figure 3.12 shows what happens when the length of the solenoid increases. Our expectation—based on our experience in introductory physics courses—is that the field should show less and less sensitivity to x as the length of the solenoid increases. And this expectation is borne out in the figure. Barring the fluctuation in the last diagram—caused by flaws in numerical calculations—even for moderate lengths of 8 and 10 (only 4 and 5 times the diameter of the solenoid), B_z appears to be fairly constant for a good fraction of the distance from the axis to the lateral surface.

The elementary treatment of a long solenoid tells us that the z-component (the only nonvanishing component) of the magnetic field should remain constant along the z-direction as well. Thus, if we plot B_z as a function of z, it should remain fairly constant inside the solenoid. The insensitivity of the field to both x and z can best be exhibited in a three-dimensional plot of B_z as a function of x and z where the field should appear as a flat sheet (except for values of x and z nearing the edges of the solenoid). Such a three-dimensional plot can be produced by typing in

```
In[10]:= Plot3D[Bz[x,0,z,a,b],{x,0,0.9},{z,-3,3}]
```

The result is Figure 3.13. Notice that, aside from the rather strong variation at the lateral surface (corresponding to x close to 1), B_z is relatively constant in both the x- and z-directions.

3.4.4 Rotating Charged Spherical Shell

Any charge in motion produces a magnetic field. This motion can be in the form of a current generated by a battery or caused by a mechanical agent acting on otherwise static charges. The general formula for the latter kind of magnetic field is

Biot–Savart law for moving charges

$$\mathbf{B} = k_m \iint_\Omega \frac{dq(\mathbf{r}')\mathbf{v}(\mathbf{r}') \times (\mathbf{r} - \mathbf{r}')}{|\mathbf{r} - \mathbf{r}'|^{3/2}} \tag{3.25}$$

where $dq(\mathbf{r}')$ is the element of charge at $\mathbf{r}'$ and $\mathbf{v}(\mathbf{r}')$ its velocity.

We are interested in the magnetic field generated by a uniformly charged spherical shell of radius a spinning about one of its diameters with an angular speed of ω. Because of the spherical geometry of the source, we use spherical coordinates for the variables of integration. Then we have,

$$\mathbf{r} = r\hat{\mathbf{e}}_r, \quad \mathbf{r}' = a\hat{\mathbf{e}}_{r'}, \quad \mathbf{v} = \omega a \sin\theta' \hat{\mathbf{e}}_{\varphi'}, \quad dq(\mathbf{r}') = \sigma a^2 \sin\theta' \, d\theta' \, d\varphi'$$

For *Mathematica* manipulations, it is convenient to express all the unit vectors in terms of Cartesian unit vectors *with coefficients in spherical*

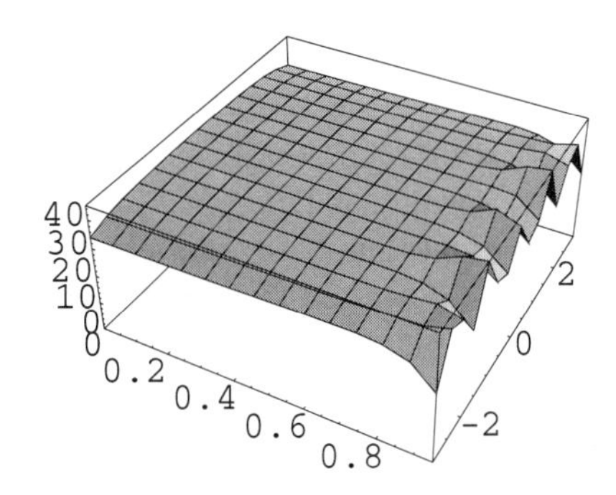

FIGURE 3.13. The z-component of the magnetic field inside a solenoid of radius 1, drawn as a function of x (distance away from the axis) and z (distance from the center on the axis).

coordinates. These are

$$\begin{aligned}
\hat{\mathbf{e}}_r &= \hat{\mathbf{e}}_x \sin\theta\cos\varphi + \hat{\mathbf{e}}_y \sin\theta\sin\varphi + \hat{\mathbf{e}}_z \cos\theta \\
\hat{\mathbf{e}}_\theta &= \hat{\mathbf{e}}_x \cos\theta\cos\varphi + \hat{\mathbf{e}}_y \cos\theta\sin\varphi - \hat{\mathbf{e}}_z \sin\theta \\
\hat{\mathbf{e}}_\varphi &= -\hat{\mathbf{e}}_x \sin\varphi + \hat{\mathbf{e}}_y \cos\varphi
\end{aligned}$$

for the field point and similar expressions (with prime on the coordinates) for the source point. The reason for doing this is that *Mathematica* treats vectors as having *Cartesian components.* For instance, it adds components to find the components of the sum, something that is not generally allowed in the so-called "curvilinear coordinates."

See *MM*, Section 1.3, especially p. 19 for components of vectors in various coordinate systems.

Because of the symmetry of the problem, we do not expect the magnetic field to depend on φ of the field point. So, for simplicity, we take φ to be zero, i.e., we position the field point in the xz-plane. We can now write the relevant code for *Mathematica.* First, we define the spherical unit vectors at the field point (with $\varphi = 0$):

```
In[1]:= er[t_]:={Sin[t],0,Cos[t]};
        et[t_]:={Cos[t],0,-Sin[t]}; ep:={0,1,0}
```

where we used t for θ and p for φ.[7] For the source point we use 1 instead of prime:

```
In[2]:= er1[t1_,p1_]:={Sin[t1]Cos[p1],Sin[t1]Sin[p1],
        Cos[t1]};
        et1[t1_,p1_]:={Cos[t1]Cos[p1],Cos[t1]Sin[p1],
        -Sin[t1]};
        ep1[p1_]:={-Sin[p1],Cos[p1],0}
```

[7]Using the BasicInput palette, one can use Greek letters for variables.

We also need the Cartesian unit vectors to find the Cartesian components of the field:

```
In[3]:= ex={1,0,0};ey={0,1,0};ez={0,0,1};
```

The velocity and the entire integrand can now be typed in:

```
In[4]:= v[t1_,p1_,a_]:= a Sin[t1] ep1[p1];
        int[r_,t_,t1_,p1_,a_] := (a^2 Sin[t1]
        Cross[v[t1,p1,a],(r er[t]-a er1[t1,p1])])
        /((r er[t]-a er1[t1,p1]).(r er[t]-
        a er1[t1,p1]))^(3/2);
```

For different components of the field, we need the corresponding components of the integrand:

```
In[5]:= intx[r_,t_,t1_,p1_,a_]:=int[r,t,t1,p1,a].ex;
        inty[r_,t_,t1_,p1_,a_]:=int[r,t,t1,p1,a].ey;
        intz[r_,t_,t1_,p1_,a_]:=int[r,t,t1,p1,a].ez;
```

Finally, we numerically integrate these integrands to obtain the components of the field:

```
In[6]:= Bx[r_,t_,a_]:=NIntegrate[Evaluate[intx[r,t,t1,
        p1,a]],{t1,0,Pi},{p1,0,2Pi}];
        By[r_,t_,a_]:=NIntegrate[Evaluate[inty[r,t,t1,
        p1,a]],{t1,0,Pi},{p1,0,2Pi}];
        Bz[r_,t_,a_]:=NIntegrate[Evaluate[intz[r,t,t1,
        p1,a]],{t1,0,Pi},{p1,0,2Pi}]
```

where we set all the nongeometrical constants multiplying the integral equal to unity.

If we type in `Bx[0,0,1]`—for the x-component of the magnetic field at the center of a sphere of radius 1—we get 9.67208×10^{-18}, indicating that $B_x = 0$ at the center. Similarly, we get zero for B_y. If we change the field point, we keep getting small numbers for B_x (and B_y) as long as we confine ourselves to the interior points of the sphere. For instance, `Bx[0.5,1.5,1]` produces -4.80235×10^{-8} and `Bx[0.25,1,1]` produces 2.3899×10^{-10}. *The x- and y-components of the magnetic field of the rotating spherical shell vanish inside.* The z-component is nonzero, though, and a plot of B_z as a function of r for $\theta = 0$ is shown in Figure 3.14. r is allowed to vary between 0 and 2. The graph on the left, produced by typing in

```
In[7]:= Bx[r_,t_,a_]:=Plot[Bz[r,0,1],{r,0,2}]
```

shows the result of calculation in detail indicating that the calculated value of B_z is very close to 8.37758, but not exactly the same for all points inside. The diagram on the right shows all values of B_z including those outside. It is obtained by typing in

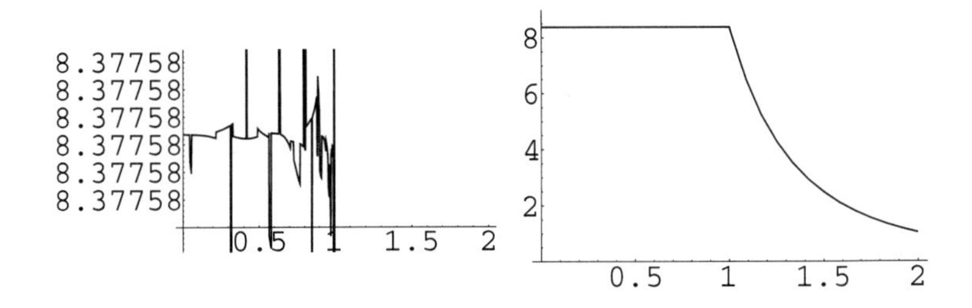

FIGURE 3.14. The z-component of the magnetic field inside a uniformly charged and uniformly spinning spherical shell of radius 1, drawn as a function of r for $\theta = 0$.

```
In[8]:= Bx[r_,t_,a_]:=Plot[Bz[r,0,1],{r,0,2},
        PlotRange->{0,9}]
```

The horizontal line for r between 0 and 1 indicates that B_z is constant inside—at least for points on the polar axis ($\theta = 0$). A sample of values of B_z for other points inside indicates that it is constant.[8] For instance,

```
Bz[0.4,Pi/4,1], Bz[0.6,Pi/3,1], Bz[0.3,Pi/6,1],
  Bz[0.9,Pi/2,1], Bz[0.8,3Pi/4,1], Bz[0.7,Pi,1]
```

all produce 8.33758. For points outside, the field drops to zero as the field point moves farther and farther away from the sphere. This behavior is evident in the diagram on the right in Figure 3.14.

3.4.5 Rotating Charged Hollow Cylinder

As the last example of magnetism (and of this chapter), we consider a uniformly charged hollow cylindrical shell of radius a and length L that is rotating with constant angular speed of ω. Once again we use Equation (3.25); however, the appropriate coordinate system is cylindrical this time. Thus,

$$\mathbf{r} = \rho\hat{\mathbf{e}}_\rho + z\hat{\mathbf{e}}_z, \quad \mathbf{r}' = a\hat{\mathbf{e}}_{\rho'} + z'\hat{\mathbf{e}}_z, \quad \mathbf{v} = \omega a\hat{\mathbf{e}}_{\varphi'}, \quad dq(\mathbf{r}') = a\, d\varphi'\, dz'$$

and we express all the unit vectors in terms of Cartesian unit vectors *with coefficients in cylindrical coordinates.* The Cartesian and cylindrical coordinates share the same $\hat{\mathbf{e}}_z$. The other two unit vectors are related as follows (see *MM*, p. 33):

$$\begin{aligned}\hat{\mathbf{e}}_\rho &= \hat{\mathbf{e}}_x \cos\varphi + \hat{\mathbf{e}}_y \sin\varphi \\ \hat{\mathbf{e}}_\varphi &= -\hat{\mathbf{e}}_x \sin\varphi + \hat{\mathbf{e}}_y \cos\varphi\end{aligned}$$

[8]It takes *Mathematica* much longer to plot B_z for nonzero values of θ. So, after a few trials, we decided not to include such plots.

As in the case of the rotating sphere, the azimuthal symmetry of the cylindrical shell prevents the magnetic field to depend on φ of the field point. So, once again, we take φ to be zero. The relevant code for *Mathematica* starts with defining the cylindrical unit vectors at the field point (with $\varphi = 0$):

```
In[1]:= er:={1,0,0}; ep:={0,1,0}
```

Again for the source point we use 1 instead of prime:

```
In[2]:= er1[p1_]:={Cos[p1],Sin[p1],0};
        ep1[p1_]:={-Sin[p1],Cos[p1],0}
```

and define the Cartesian unit vectors as usual:

```
In[3]:= ex={1,0,0};ey={0,1,0};ez={0,0,1};
```

The velocity and the entire integrand can now be typed in:

```
In[4]:= v[p1_,a_]:= a ep1[p1];
        int[r_,z_,p1_,z1_,a_]:= (a Cross[v[z1,p1,a],
        (r er-a er1[p1]+(z-z1) ez)]) /((r er-a er1[p1]
        +(z-z1) ez).(r er-a er1[p1]+(z-z1) ez))|^(3/2);
```

For different components of the field, we need the corresponding components of the integrand:

```
In[5]:= intx[r_,z_,p1_,z1_,a_]:=int[r,z,p1,z1,a].ex;
        inty[r_,z_,p1_,z1_,a_]:=int[r,z,p1,z1,a].ey;
        intz[r_,z_,p1_,z1_,a_]:=int[r,z,p1,z1,a].ez;
```

Finally, as before, we set all the nongeometrical constants multiplying the integral equal to unity, and numerically integrate the integrands to obtain the components of the field:

```
In[6]:= Bx[r_,z_,L_,a_]:=NIntegrate[Evaluate[intx[r,z,p1,
        z1,a]],{z1,-L/2,L/2},{p1,0,2Pi}];
        By[r_,z_,L_,a_]:=NIntegrate[Evaluate[inty[r,z,p1,
        z1,a]],{z1,-L/2,L/2},{p1,0,2Pi}];
        Bz[r_,z_,L_,a_]:=NIntegrate[Evaluate[intz[r,z,p1,
        z1,a]],{z1,-L/2,L/2},{p1,0,2Pi}]
```

We are interested in the behavior of B_z as the field point moves in the xy-plane. This could be done by using a command such as

```
Plot[Bz[r,0,10,1],{r,0,1.5}]
```

for a cylinder of unit radius and length $L = 10$. However, because of the small values of the integral outside, *Mathematica* would require a long time to do the plot. The judicious alternative is to make a table at selected values of r, and, if desired, plot the table using `ListPlot`.

r[i]	c[i]	b[i]	r[i]	c[i]	b[i]
0.0	3.04779	12.5563	1.01	−4.4053	−0.00997618
0.2	3.13132	12.5563	1.15	−2.59646	−0.0100094
0.5	3.6704	12.5563	1.2	−2.13725	−0.0100065
0.8	5.3922	12.5563	1.25	−1.7705	−0.0100036
0.9	6.55391	12.5563	1.3	−1.47953	−0.0100006
0.95	7.26212	12.5564	1.35	−1.24818	−0.00999741
0.99	7.86082	12.5563	1.45	−0.913366	−0.00999073

TABLE 3.1. Values (`c[i]`) of the magnetic field of a short cylinder and a long cylinder (`b[i]`).

The first cylinder has unit radius, length 0.5, and its magnetic field is denoted by `c[i]`. To get the fifth value, for example, of B_z—corresponding to `r[5]` (which is 0.9)—for this cylinder, one types in

```
c[5]=Bz[0.9,0,0.5,1]
```

The second cylinder has unit radius, length 50, and its magnetic field is denoted by `b[i]`. To get the eighth value of B_z—corresponding to `r[8]` (which is 1.01)—for this cylinder, one types in

```
b[8]=Bz[1.01,0,50,1]
```

In separate inputs, we assign values to r. For example,

```
r[1]=0; r[2]=0.2; r[3]=0.5; r[4]=0.8;
```

and similar inputs for `r[5]` through `r[14]`.

To construct the table of values, we type in

```
Table[{r[i], c[i], b[i]}, {i, 1, 15}] // MatrixForm
```

and *Mathematica* produces a three-column table, which we have reproduced (in six-column format) in Table 3.1. It is clear that the latter has the characteristics of a long solenoid, i.e., constant magnetic field inside (the left half of the table) and—almost—zero magnetic field outside (the right half of the table).

3.5 Problems

Problem 3.1. A segment of the parabola $y = x^2$ extending from $x = -1$ to $x = 1$ has a uniform linear charge density.

(a) Write a *Mathematica* code to calculate the electrostatic potential of this charge distribution.

(b) Plot the potential for

(i) points in the xy-plane, and
(ii) points in the plane $z = 0.5$ parallel to the xy-plane.
Hint: For the parameter of the curve, choose $x = t$, i.e., let $f(t) = t$, then find $g(t)$.

Problem 3.2. Consider a uniform linear charge distribution in the form of an ellipse with a semimajor axis equal to 6 and a semiminor axis equal to 1. Use Cartesian coordinates and the parametric equation of the ellipse in terms of trigonometric functions.
(a) Write down the single integral that gives the electric potential at an arbitrary point P in space.
(b) Specialize to the case where P lies in the plane of the ellipse. Plot the resulting potential.
(c) Specialize to the case where P lies in the plane $z = 0.5$ parallel to the xy-plane. Plot the resulting potential.

Problem 3.3. Using Equation (3.16), plot the potential of the ring of Section 3.3.1 as a function of x and y for points in the plane $z = 0.25$ parallel to the xy-plane.

Problem 3.4. Using Equation (3.15), plot the x-component of the field of the ring of Section 3.3.1 as a function of x and y for points in the xy-plane.

Problem 3.5. Using Equation (3.15), plot the y-component of the field of the spiral of Section 3.3.2 as a function of x and y for points in the xy-plane.

Problem 3.6. Electric charge is distributed uniformly on a thin straight wire of unit length lying along the x-axis of a Cartesian coordinate system with origin at its midpoint. Plot the electrostatic potential at points on a plane parallel to the z-axis whose distance from xy-plane is 0.25. Hint: Choose x' to be the parameter of the curve; i.e., let $f(t) = t$.

Problem 3.7. Electric charge is distributed uniformly on a thin wire whose equation in polar coordinates is $r = 3 - 2\cos\theta$.
(a) Construct the figure of this curve by writing its equation in Cartesian coordinates and using `ParametricPlot`.
(b) Plot the electrostatic potential Φ at points on a plane parallel to the z-axis whose distance from xy-plane is 0.5.
(c) Use Equation (3.15) to write the components of the electric field at points on the plane of part (b) in terms of integrals.
(d) Write three *Mathematica* expressions defining the three components of the field—as functions of x and y—in terms of integrals.
(e) Write another *Mathematica* expression defining the *absolute value* of the field—again as a function of x and y—in terms of the components.
(f) Plot the absolute value of the field at points on the plane of (b) as a function of x and y.

Problem 3.8. Electric charge is distributed uniformly on a square surface of side 1. Use Equation (3.22) to plot the potential of this charge distribution for points of the plane $z = 0.005$ parallel to the xy-plane. Restrict the plot to $-1 \le x \le 1$ and $-1 \le y \le 1$.

Problem 3.9. Electric charge is distributed uniformly on a square surface of side 1. Use Equation (3.20) to plot the magnitude of the electric field of this charge distribution for points of the plane $z = 0.005$ parallel to the xy-plane. Restrict the plot to $-1 \le x \le 1$ and $-1 \le y \le 1$. Hint: You can do one of the integrals of the double integration analytically.

Problem 3.10. Electric charge is distributed uniformly on a circular disk of diameter 1. Use Equation (3.21) to plot the potential of this charge distribution for points of the plane $z = 0.005$ parallel to the xy-plane. Restrict the plot to $-1 \le x \le 1$ and $-1 \le y \le 1$. Hint: You can do one of the integrals of the double integration analytically.

Problem 3.11. Electric charge is distributed uniformly on a circular disk of diameter 1. Use Equation (3.20) to plot the magnitude of the electric field of this charge distribution for points of the plane $z = 0.005$ parallel to the xy-plane. Restrict the plot to $-1 \le x \le 1$ and $-1 \le y \le 1$. Hint: You can do one of the integrals of the double integration analytically.

Problem 3.12. Electric charge is distributed uniformly on an elliptical disk of semimajor (along the x-axis) and semiminor (along the y-axis) axes 5 and 1, respectively. Use Equation (3.21) to plot the potential of this charge distribution for points of the plane $z = 0.005$ parallel to the xy-plane. Restrict the plot to $-6 \le x \le 6$ and $-2 \le y \le 2$. Hint: You can do one of the integrals of the double integration analytically.

Problem 3.13. Electric charge is distributed uniformly on an elliptical disk of semimajor (along the x-axis) and semiminor (along the y-axis) axes 5 and 1, respectively. Use Equation (3.20) to plot the magnitude of the electric field of this charge distribution for points of the plane $z = 0.005$ parallel to the xy-plane. Restrict the plot to $-6 \le x \le 6$ and $-2 \le y \le 2$. Hint: You can do one of the integrals of the double integration analytically.

Problem 3.14. Using the results of Section 3.4.2, find the *magnitude* of the magnetic field of a current loop in the shape of an ellipse with a semimajor axis equal to 8 and a semiminor axis equal to 1, and plot it as a function of x and y for points on a plane $z = 0.5$ parallel to the xy-plane. Hint: Find the parametric equation of the ellipse in terms of trigonometric functions.

Problem 3.15. Using the results of Section 3.4.2, find the *magnitude* of the magnetic field of a current loop in the shape of a helix with elliptical cross section with a semimajor axis equal to 5 and a semiminor axis equal to 1. Assume that the lower end of the helix is in the xy-plane, the distance between windings is 0.1, and the number of windings is 10. Plot the magnitude of the field as a function of x and y for points on a plane $z = -0.25$ parallel to the xy-plane.

4
Infinite Series and Finite Sums

The laws of physics are exact expressions of an inexact universe. They are translations of concrete observations into the ideal language of mathematics. This kinship between the concrete and the ideal is the highest intellectual achievement of mankind and the driving force behind our ability to unravel the secrets of the universe. However, the laws, as they stand, are incapable of describing the objects of the universe, as *they* stand. Therefore, one has to *approximate* the objects of the universe in such a way that the laws can be applied to them. One method of approximation uses infinite series, which are best studied in the context of infinite sequences.

4.1 Infinite Sequences

An infinite sequence is like a function, except that instead of real numbers, its domain is the set of positive integers. In *Mathematica* we write sequences in exactly the same way as functions. For example,

```
In[1]:= a[i_]:=(1+1/i)^i
```

represents the sequence

MM, pp. 205–208

$$\{2, 1.5^2, (1+\tfrac{1}{3})^3, \dots\} \quad \text{or} \quad \left\{\left(1+\frac{1}{k}\right)^k\right\}_{k=1}^{\infty}$$

An important sequence is the sequence of **partial sums**. As the name suggests, it is a sequence whose members are sums of an increasing number of terms. For instance, if we define s_n as[1]

$$s_n = \sum_{k=0}^{n} \frac{(-1)^k}{k+1}$$

then we can construct a sequence—the sequence of partial sums—whose members are s_0, s_1, s_2, etc. An explicit enumeration of this sequence would be

$$\{1, 1-\tfrac{1}{2}, 1-\tfrac{1}{2}+\tfrac{1}{3}, 1-\tfrac{1}{2}+\tfrac{1}{3}-\tfrac{1}{4}, \dots\}$$

In *Mathematica* this sequence could be defined as

```
In[2]:= s[n_]:=Sum[(-1)^k/(k+1),{k,0,n}]
```

We can tell *Mathematica* to display the sequence up to a certain number of elements by using the `Table` command. Thus, for `a[i]` defined above

```
In[3]:= t=Table[a[i],{i,1,5}];
```

creates a list, which we have named `t`, consisting of a_1 through a_5. If we type `t`, *Mathematica* puts out

$$\left\{2, \frac{4}{9}, \frac{64}{27}, \frac{625}{256}, \frac{7776}{3125}\right\}$$

Similarly,

```
In[3]:= Table[s[i],{i,0,5}];
```

puts out

$$\left\{1, \frac{1}{2}, \frac{5}{6}, \frac{7}{12}, \frac{47}{60}, \frac{37}{60}\right\}$$

The most important property of a sequence is the nature of its *convergence*. This is particularly important in the case of the partial sums, because the nature of convergence tells us whether the corresponding *infinite series* makes sense. The limit of a sequence is the ultimate number obtained—if any—as the subscript (or the argument) gets larger and larger. In *Mathematica* this is done using `Limit`. For example,

use of **Limit** for sequences in *Mathematica*

```
In[4]:= Limit[a[i],i->Infinity]
```

produces

```
Out[4]:= E
```

However,

[1]It is common in mathematics literature to distinguish members of a sequence by their subscripts rather than arguments (as in functions).

```
In[5]:= Limit[s[i],i->Infinity]
```

yields

$$\text{Limit}\left[\frac{\text{Log}[4]}{2} - \frac{1}{2}(-1)^n \left(\text{PolyGamma}\left[0, \frac{2+n}{2}\right] - \text{PolyGamma}\left[0, \frac{3+n}{2}\right]\right), n \to \infty\right]$$

the first term of which is equivalent to $\ln(2)$, the known value of the infinite series. The remaining term is really zero, but the internal routine of taking limits does not evaluate it to zero. However, if we type in

MM, pp. 219 and 225

```
In[6]:= s[Infinity]
```

then *Mathematica*'s output will consist of only $\frac{\text{Log}[4]}{2}$.

It turns out that *Mathematica* recognizes many infinite series, although its output may appear strange. Suppose we construct a sequence of partial sums by typing in

```
In[7]:= b[n_]:=Sum[1/k^2,{k,1,n}]
```

and tell *Mathematica* to evaluate the infinite series corresponding to this partial sum:

```
In[8]:= b[Infinity]
```

The result is put out instantly as $\pi^2/6$. In fact, *Mathematica* recognizes a more general infinite series. Instead of the partial sum above, let us define

```
In[9]:= z[p_,n_]:=Sum[1/k^p,{k,1,n}]
```

and then type in

```
In[10]:= z[p,Infinity]
```

Mathematica's output will be `Zeta[p]`. The series turns out to be a famous series in mathematical physics known as the **Riemann zeta function** denoted by $\zeta(p)$, where p can be any real (or complex) number. For even p, the zeta function evaluates to a multiple of π^p. For example, `z[34,Infinity]` (or `Zeta[34]`) generates the output

Riemann zeta function

$$\frac{151628697551\pi^{34}}{12130454581433748587292890625}$$

However, `z[33,Infinity]` (or `Zeta[33]`) echoes the input. If we are interested in the *approximation* to the series, then we can insert the expression in `N[ ]`. For example,

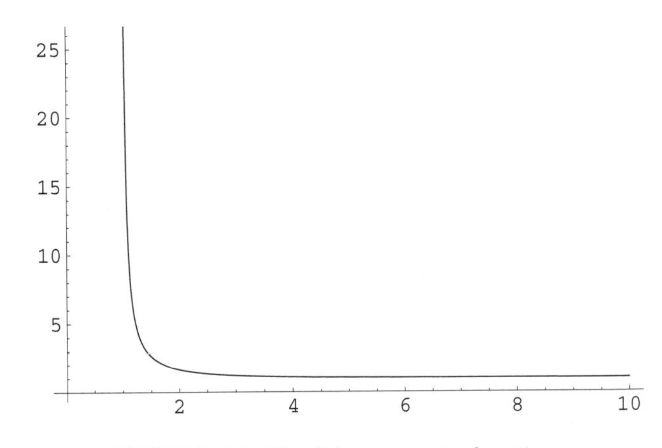

FIGURE 4.1. The Riemann zeta function.

```
In[11]:= N[z[33,Infinity],15]
```

gives 1.00000000011642. It is clear from the definition of the zeta function that $\zeta(p)$ approaches 1 as p goes to infinity. Furthermore, $\zeta(1)$ is infinite, because the corresponding series can be shown to diverge. Figure 4.1, obtained by typing in

```
In[12]:= Plot[Zeta[p],{p,1,10}]
```

MM, pp. 215–216

summarizes the behavior of the Riemann zeta function.

MM, p. 231

Although *Mathematica* recognizes many infinite series, it does so in an ostensibly unconventional way. For instance, the well-known series

$$\sum_{k=0}^{\infty} \frac{(-1)^k}{(2k+1)!}$$

when typed in as

```
In[13]:= c[n_]:=Sum[(-1)^k/(2k+1)!,{k,0,n}]; c[Infinity]
```

produces

$$\sqrt{\frac{\pi}{2}}\ \text{BesselJ}\left[\frac{1}{2}, 1\right]$$

MM, p. 283

instead of $\sin(1)$. Nevertheless, the two are completely equivalent.

4.2 Numerical Integration

The recurrence of integration in all aspects of mathematical physics makes the techniques of numerical integration worthy of careful study. Such a study involves using sums, and since this chapter discusses the concept of sum and series, it is worth our time to take a brief detour into numerical integration.

Recall that the integral of a function f is the limit of a sum in which each term is the product of the width of a small interval and the value of f at a point in that interval, i.e., the sum of the areas of (a large number of) rectangles. We start with this notion and later find approximations to integrals that are much more accurate than sums of rectangular areas.

The basic idea is to write

$$\int_a^{a+\Delta x} f(x)\,dx \approx \sum_{i=0}^{N} W_i f_i \tag{4.1}$$

where Δx is usually—but not necessarily—small, N is the number of intervals in $(a, a+\Delta x)$, f_i is $f(x_i)$ with x_i the "evaluation" point in the ith interval, and W_i is a "weight factor" to be determined for convenience and accuracy. It is desirable to space all evaluation points equally and call the spacing h. Depending on the number of intervals chosen, one obtains various approximation formulas for the integral.

The general procedure is to choose N and demand that Equation (4.1) hold exactly when f is *any* arbitrary polynomial of degree N. This is equivalent to demanding that the equality hold for each power of x up to N.

4.2.1 The Simplest Method

As the simplest case, assume that $N = 0$. Then the only restriction is

$$\int_{x_0}^{x_1} 1\,dx = W_0 \;\Rightarrow\; x_1 - x_0 = W_0 \quad \text{or} \quad W_0 = \Delta x = h$$

with $x_0 = a$ and $x_1 = a + \Delta x = a + h$. The approximation to the integral is

$$\int_a^{a+\Delta x} f(x)\,dx \approx W_0 f_0 \equiv h f_0$$

i.e., the area of the rectangle shown in Figure 4.2(a).

It is clear that the approximation is very crude if Δx is large and f varies a great deal between a and $a + \Delta x$. To remedy this situation one divides the interval $(a, a+\Delta x)$ into a large number of subintervals and for each subinterval the approximation above is applied. More specifically, consider the integral of f over the interval (a, b). Divide the interval (a, b) into n subintervals each of length $(b-a)/n$, which we call h.

Because of the smallness of the subintervals, each rectangle approximates the area under the subinterval rather well. So, the sum of the areas of the

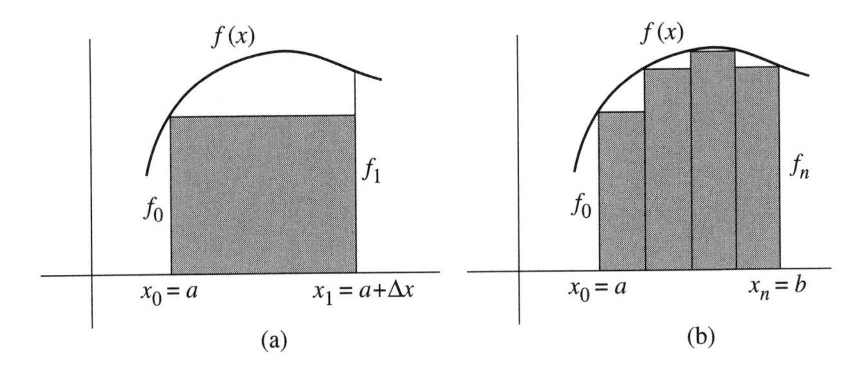

FIGURE 4.2. (a) The simplest rule for numerical integration approximates the area under a curve with that of a rectangle. (b) A better approximation is obtained if the region of integration is divided into smaller subintervals.

small rectangles is a much better approximation to the true area than the single rectangle [Figure 4.2(b)]. This sum can be explicitly calculated:

$$\int_a^b f(x)\,dx = \int_a^{a+h} f(x)\,dx + \int_{a+h}^{a+2h} f(x)\,dx + \cdots + \int_{a+(n-1)h}^{a+nh} f(x)\,dx$$
$$\approx hf_0 + hf_1 + \cdots + hf_n = h(f_0 + f_1 + \cdots + f_n) \tag{4.2}$$

where $f_k = f(x_k) = f(a + kh)$ and $x_n = a + nh = b$.

To illustrate the discussion above, let us apply the method to the integral $\int_0^1 e^x\,dx$, whose value we know to be $e - 1$. The *Mathematica* code for this is very simple:

```
In[1]:=h[n_]:=1./n;intRec[n_]:=h[n] Sum[E^(k h[n]),{k,0,n}]
```

We get a better and better approximation to the value of the integral as we make n larger and larger. Thus, `intRec[100]` gives 1.73689 and `intRec[10000]` gives 1.71848 compared to 1.7182818285, which is the exact result to 11 significant figures. It is clear that this simple numerical method of integration is not very accurate.

4.2.2 Trapezoid Rule

For better accuracy, we need to increase N. So now let $N = 1$ and demand that the approximation in (4.1) be exact when $f(x) = 1$ and $f(x) = x$. This leads to two equations:

$$\int_{x_0}^{x_1} 1\,dx = W_0 \cdot 1 + W_1 \cdot 1 \;\Rightarrow\; x_1 - x_0 = W_0 + W_1$$
$$\int_{x_0}^{x_1} x\,dx = W_0 x_0 + W_1 x_1 \;\Rightarrow\; \frac{x_1^2 - x_0^2}{2} = W_0 x_0 + W_1 x_1$$

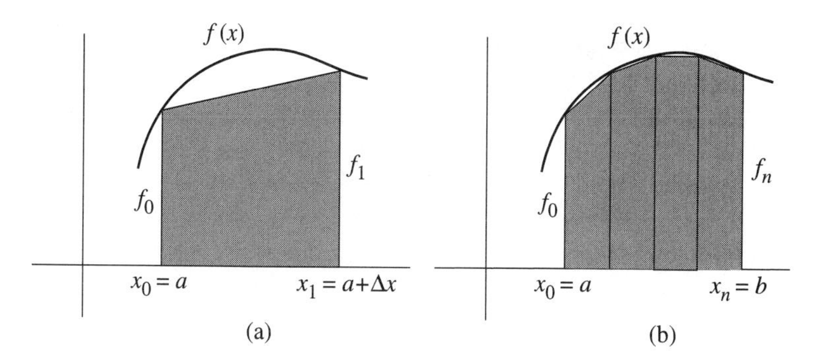

FIGURE 4.3. (a) The trapezoid rule approximates the area under a curve with that of a trapezoid obtained by replacing the curve with a straight line. (b) To obtain accuracy, the interval of integration should be subdivided.

whose solution is easily found to be

$$W_0 = W_1 = \frac{x_1 - x_0}{2} = \frac{\Delta x}{2} \equiv \frac{h}{2} \tag{4.3}$$

The integral itself is then

trapezoid rule

$$\int_a^{a+\Delta x} f(x)\,dx = \int_a^{a+h} f(x)\,dx \approx \frac{h}{2}(f_0 + f_1) = \frac{h}{2}[f(a) + f(a+h)] \tag{4.4}$$

This is called the **trapezoid rule** because we have effectively replaced the area under a curve by the area under the trapezoid[2] obtained by approximating the curve with a straight line as shown in Figure 4.3(a).

To find the numerical value of the integral $\int_a^b f(x)\,dx$, we divide the interval (a, b) into n subintervals as before and let $(b-a)/n = h$ [Figure 4.3(b)]. The sum of the areas of the small trapezoids can now be explicitly calculated:

$$\begin{aligned}\int_a^b f(x)\,dx &= \int_a^{a+h} f(x)\,dx + \int_{a+h}^{a+2h} f(x)\,dx + \cdots + \int_{a+(n-1)h}^{a+nh} f(x)\,dx \\ &\approx \frac{h}{2}(f_0 + f_1) + \frac{h}{2}(f_1 + f_2) + \cdots + \frac{h}{2}(f_{n-1} + f_n)\end{aligned}$$

where again $f_k = f(x_k) = f(a + kh)$ and $x_n = a + nh = b$. It now follows that

$$\int_a^b f(x)\,dx \approx h\left(\frac{f_0}{2} + f_1 + f_2 + \cdots + f_{n-1} + \frac{f_n}{2}\right) \tag{4.5}$$

[2]The reader may recall that the area of a trapezoid is the product of the sum of its parallel sides and the perpendicular distance between those two sides.

It is interesting to note that the difference between the trapezoid rule of (4.5) and the rectangular method of (4.2) is only the two factors of $\frac{1}{2}$ multiplying f_0 and f_n. This "small" change, however, makes a tremendous difference in the outcome, as we illustrate shortly.

Let us apply the method to our integral $\int_0^1 e^x\,dx$. The *Mathematica* code for this is easily written:

```
In[2]:= h[n_]:=1./n; intTrap[n_]:=(h[n]/2)(1+E^(n h[n]))
        +h[n] Sum[E^(k h[n]),{k,1,n-1}]
```

Typing in `N[intTrap[100],10]` gives 1.718296147, which is a great improvement over the previous method.

4.2.3 Simpson's Rule

If we want an improvement over the trapezoid rule, we let $N = 2$ in Equation (4.1) and demand that the approximation give exactly the same result as the integral for 1, x, and x^2. Then, with $x_0 = a$, $x_1 = a + h$, and $x_2 = a + \Delta x = a + 2h$, we obtain

$$\int_{x_0}^{x_2} 1\,dx = W_0 \cdot 1 + W_1 \cdot 1 + W_2 \cdot 1 \;\Rightarrow\; x_2 - x_0 = W_0 + W_1 + W_2$$

$$\int_{x_0}^{x_2} x\,dx = W_0 x_0 + W_1 x_1 + W_2 x_2 \;\Rightarrow\; \frac{x_2^2 - x_0^2}{2} = W_0 x_0 + W_1 x_1 + W_2 x_2$$

$$\int_{x_0}^{x_2} x^2\,dx = W_0 x_0^2 + W_1 x_1^2 + W_2 x_2^2 \;\Rightarrow\; \frac{x_2^3 - x_0^3}{3} = W_0 x_0^2 + W_1 x_1^2 + W_2 x_2^2$$

for whose solution we use *Mathematica* and type in

```
In[1]:= eqn1=2h==W0+W1+W2;
        eqn2=((x0+2 h)^2-x0^2)/2==W0 x0+W1 (x0+h)
        +W2 (x0+2 h);
        eqn3=((x0+2 h)^3-x0^3)/3==W0 x0^2+W1 (x0+h)^2
        +W2 (x0+2 h)^2;
```

and

```
In[2]:= Solve[{eqn1,eqn2,eqn3},{W0,W1,W2}]
```

Mathematica then gives the following solution instantly:

$$\left\{\left\{W_0 \to \frac{h}{3}, \quad W_1 \to \frac{4h}{3}, \quad W_2 \to \frac{h}{3}\right\}\right\}$$

Simpson's rule

We thus obtain **Simpson's rule**:

$$\int_a^{a+2h} f(x)\,dx \approx \frac{h}{3}(f_0 + 4f_1 + f_2) \equiv \frac{h}{3}[f(a) + 4f(a+h) + f(a+2h)] \tag{4.6}$$

Simpson's rule becomes useful only if h is small. This is achieved, as before, by dividing the interval (a, b) into a large (even) number n of subintervals on each pair of which we apply the rule. It then follows that

$$\int_a^b f(x)\,dx = \int_a^{a+2h} f(x)\,dx + \int_{a+2h}^{a+4h} f(x)\,dx + \cdots + \int_{a+(n-2)h}^{a+nh} f(x)\,dx$$
$$\approx \frac{h}{3}(f_0 + 4f_1 + f_2) + \frac{h}{3}(f_2 + 4f_3 + f_4) + \cdots + \frac{h}{3}(f_{n-2} + 4f_{n-1} + f_n)$$

or

$$\int_a^b f(x)\,dx \approx \frac{h}{3}\left(f_0 + 4f_1 + 2f_2 + 4f_3 + \cdots + 2f_{n-2} + 4f_{n-1} + f_n\right) \quad (4.7)$$

where as before $f_k = f(x_k) = f(a + kh)$ and $x_n = a + nh = b$.

Once again, let us apply Simpson's method to our integral $\int_0^1 e^x\,dx$. The *Mathematica* code is only slightly more complicated:

```
In[3]:= h[n_]:=1./n; intSimp[n_]:=(h[n]/3)(1+E^(n h[n]))
        +(4 h[n]/3)Sum[E^((2 k+1) h[n]),{k,0,(n-2)/2}]
        +(2 h[n]/3)Sum[E^(2 k h[n]),{k,0,(n-2)/2}]
```

Typing in `intSimp[10]` gives 1.71828—a great improvement over the trapezoid method—and N[`intSimp[100],10`] yields 1.7182818286, different from the "exact" result only at the 11th significant figure.

We can continue this general procedure to obtain more and more accurate results. Thus, with $N = 3$, we obtain

$$I \approx W_0 f_0 + W_1 f_1 + W_2 f_2 + W_3 f_3$$

which we demand to hold exactly for $f = 1$, $f = x$, $f = x^2$, and $f = x^3$. This yields four equations in four unknowns whose solution is found to be

$$W_0 = W_3 = \frac{3h}{8}, \qquad W_1 = W_2 = \frac{9h}{8}$$

This leads to the **Simpson's three-eighths rule**:

Simpson's three-eighths rule

$$\int_a^{a+3h} f(x)\,dx \approx \frac{3h}{8}(f_0 + 3f_1 + 3f_2 + f_3)$$

When $N = 4$, we obtain **Boole's rule**:

Boole's rule

$$\int_a^{a+4h} f(x)\,dx \approx \frac{2h}{45}(7f_0 + 32f_1 + 12f_2 + 32f_3 + 7f_4)$$

These formulas can be used to find the integral of a function by dividing the interval into a large number (a multiple of 3 for Simpson's three-eighths rule, and a multiple of 4 for Boole's rule) of subintervals and using the corresponding rule for each 3 or 4 subintervals. We leave the details to the reader. We expect each rule to be more accurate than the preceding one in the limit of smaller and smaller h.

4.2.4 Gaussian Integration

In the previous subsection, we assumed equidistant spacing h for the evaluation points. This, while convenient, is an unnecessary restriction that can prevent us from exploring the possibilities afforded by the freedom to *choose* the evaluation points. Gauss used this freedom as well as the nice properties of orthogonal polynomials to come up with a very powerful technique of evaluating integrals numerically called **Gaussian quadrature integration.**

Gaussian quadrature integration

Consider the integral

$$\int_a^b f(x)w(x)\,dx \approx \sum_{i=1}^{N} W_i f(x_i) \tag{4.8}$$

MM, pp. 193–197

where $w(x)$ is a function that is always positive in the interval (a, b), for example, $w(x) = 1$. Its presence allows us to use the machinery of orthogonal polynomials as we shall see shortly. We have to find the $2N$ unknowns in (4.8). These are the N weights W_i and the N abscissas x_i. The determination of these unknowns requires $2N$ equations obtained by demanding that Equation (4.8) hold *exactly* for polynomials of order $2N - 1$ or less. Two particular types of polynomials naturally produce simple equations from which the unknowns can be calculated.

Let $Q_{N-1}(x)$ be *any* polynomial of degree $N - 1$ and $F_N(x)$ an *orthogonal* polynomial of degree N defined on (a, b) with weight function $w(x)$. The product $Q_{N-1}(x)F_N(x)$ is a polynomial of degree $2N - 1$. So, by assumption, it must satisfy Equation (4.8) exactly:

$$\int_a^b Q_{N-1}(x)F_N(x)w(x)\,dx = \sum_{i=1}^{N} W_i Q_{N-1}(x_i)F_N(x_i)$$

Since $Q_{N-1}(x)$ is a polynomial of degree $N-1$, it can be written as a linear combination of $F_k(x)$ with $k < N$: $Q_{N-1}(x) = \sum_{k=0}^{N-1} \alpha_k F_k(x)$. Then

$$\begin{aligned}\int_a^b Q_{N-1}(x)F_N(x)w(x)\,dx &= \int_a^b \sum_{k=0}^{N-1} \alpha_k F_k(x)F_N(x)w(x)\,dx \\ &= \sum_{k=0}^{N-1} \alpha_k \underbrace{\int_a^b F_k(x)F_N(x)w(x)\,dx = 0}_{=0 \text{ because } k \neq N}\end{aligned}$$

It follows that

$$\sum_{i=1}^{N} W_i Q_{N-1}(x_i)F_N(x_i) = 0$$

For this to hold for every polynomial, we must demand $F_N(x_i) = 0$, i.e. *the evaluation points (or abscissas) of (4.8) are the roots of the Nth orthogonal polynomial appropriate for the interval (a, b) and weight function $w(x)$.*

To find W_i, we evaluate Equation (4.8) for a particular polynomial. Consider the **Lagrange's interpolating polynomial**

Lagrange's interpolating polynomial

$$l_{j,N}(x) \equiv \frac{(x-x_1)\cdots(x-x_{j-1})(x-x_{j+1})\cdots(x-x_N)}{(x_j-x_1)\cdots(x_j-x_{j-1})(x_j-x_{j+1})\cdots(x_j-x_N)} \tag{4.9}$$

that has the property that

$$l_{j,N}(x_i) = \begin{cases} 0 & \text{if} \quad i \neq j \\ 1 & \text{if} \quad i = j \end{cases}$$

This polynomial is of degree $N-1$ and, therefore, it must satisfy (4.8) exactly. This gives

$$\int_a^b l_{j,N}(x)w(x)\,dx = \sum_{i=1}^N W_i l_{j,N}(x_i) = W_j$$

Thus, *the weight factors W_i of (4.8) are given by*

$$W_i = \int_a^b l_{i,N}(x)w(x)\,dx \tag{4.10}$$

where $l_{i,N}(x)$ is Lagrange's interpolating polynomial of (4.9).

The Gaussian integration may seem restrictive due to the apparent specificity of the limits of integration and the weight function $w(x)$ for orthogonal polynomials. For example, **Gauss–Legendre** integration—in which Legendre polynomials are used—will work only if the interval of integration happens to be $(-1,+1)$ and $w(x)=1$. While the latter condition makes Legendre polynomials attractive (we don't have to worry about any weight functions), the fact that Legendre polynomials reside only in the interval $(-1,+1)$ may lead us to think that they are useless for a general interval (a,b). However, this is not the case because the simple change of variable

Gauss–Legendre integration

$$y = \frac{2}{b-a}x - \frac{b+a}{b-a} \qquad \text{or} \qquad x = \frac{b-a}{2}y + \frac{b+a}{2} \tag{4.11}$$

changes the limits of integration from (a,b) to $(-1,+1)$, as the reader may verify.

Now let us evaluate the integral $\int_a^b f(x)\,dx$ using the Gauss–Legendre integration procedure and choosing N to be 2. First we transform the variable of integration according to (4.11):

$$\begin{aligned}\int_a^b f(x)\,dx &= \int_{-1}^1 f\left(\frac{b-a}{2}y + \frac{b+a}{2}\right)\left(\frac{b-a}{2}\,dy\right) \\ &\equiv \int_{-1}^1 g(y)\,dy = \int_{-1}^1 g(x)\,dx\end{aligned}$$

where in the last step we changed the dummy variable of integration and defined g by

$$g(x) = \frac{b-a}{2} f\left(\frac{b-a}{2}x + \frac{b+a}{2}\right)$$

Next we find the evaluation points and the weight factors W_1 and W_2. The evaluation points are simply the roots of the Legendre polynomial of degree 2. Since

$$P_2(x) = \tfrac{1}{2}(3x^2 - 1)$$

we have

$$x_1 = -\frac{1}{\sqrt{3}} \quad \text{and} \quad x_2 = \frac{1}{\sqrt{3}}$$

The weight factors are obtained by using Equation (4.10), or

$$W_1 = \int_{-1}^{1} l_{1,2}(x)\, dx = \int_{-1}^{1} \frac{x - x_2}{x_1 - x_2}\, dx = -\frac{\sqrt{3}}{2}\int_{-1}^{1}\left(x - \frac{1}{\sqrt{3}}\right) dx = 1$$

$$W_2 = \int_{-1}^{1} l_{2,2}(x)\, dx = \int_{-1}^{1} \frac{x - x_1}{x_2 - x_1}\, dx = \frac{\sqrt{3}}{2}\int_{-1}^{1}\left(x + \frac{1}{\sqrt{3}}\right) dx = 1$$

Using these values yields

$$\begin{aligned}\int_a^b f(x)\, dx &= \frac{b-a}{2}\int_{-1}^{1} f\left(\frac{b-a}{2}x + \frac{b+a}{2}\right) dx \approx W_1 g(x_1) + W_2 g(x_2) \\ &= \frac{b-a}{2}\left[f\left(-\frac{b-a}{2\sqrt{3}} + \frac{b+a}{2}\right) + f\left(\frac{b-a}{2\sqrt{3}} + \frac{b+a}{2}\right)\right]\end{aligned}$$

Denoting $b - a$ by h, we obtain

$$\int_a^b f(x)\, dx \approx \frac{h}{2}\left[f\left(\frac{h}{2}\left(1 - \frac{1}{\sqrt{3}}\right) + a\right) + f\left(\frac{h}{2}\left(1 + \frac{1}{\sqrt{3}}\right) + a\right)\right] \tag{4.12}$$

Equation (4.12) can be used to evaluate integrals whose interval of integration has been divided into subintervals as in trapezoid and Simpson's rules. More specifically, suppose we are interested in the integral of $f(x)$ over the (large) interval (a, b). As usual we divide the interval into n subintervals. Let x_k denote the endpoint of the kth subinterval. Then

$$\begin{aligned}\int_a^b f(x)\, dx &= \int_a^{x_1} f(x)\, dx + \int_{x_1}^{x_2} f(x)\, dx + \cdots + \int_{x_{n-1}}^{b} f(x)\, dx \\ &= \sum_{k=0}^{n-1}\int_{x_k}^{x_{k+1}} f(x)\, dx\end{aligned}$$

where $x_0 \equiv a$ and $x_n \equiv b$. For the integral inside the sum, we can use (4.12). This leads to the following integration rule:

Gauss–Legendre integration rule

$$\int_a^b f(x)\,dx \approx \frac{h}{2}\sum_{k=0}^{n-1}\left[f\left(\frac{h}{2}\left(1-\frac{1}{\sqrt{3}}\right)+x_k\right)+f\left(\frac{h}{2}\left(1+\frac{1}{\sqrt{3}}\right)+x_k\right)\right]$$
$$= \frac{h}{2}\sum_{k=0}^{n-1}\left[f\left(\left(\frac{\sqrt{3}-1}{2\sqrt{3}}+k\right)h+a\right)+f\left(\left(\frac{\sqrt{3}+1}{2\sqrt{3}}+k\right)h+a\right)\right] \tag{4.13}$$

where we used the fact that $x_k = a + kh$.

To evaluate our integral $\int_0^1 e^x\,dx$, we write the following *Mathematica* code:

```
In[1]:= h[n_]:=1./n; intGauss[n_]:=(h[n]/2) Sum[E^(k+0.5
        -1/(2Sqrt[3]))h[n] +E^(k+0.5+1/(2Sqrt[3]))h[n],
        {k,0,n-1}]
```

Typing in `intGauss[30]` gives 1.718281828, a great improvement over Simpson's method, which yields this same result only if $n = 130$ instead of 30.

Equation (4.12) is derived by using two evaluation points in (4.8). One can think of (4.12) as analogous to the trapezoid rule. If $N = 3$ in (4.8), we obtain the analog of Simpson's rule. Since Gaussian integration (with $N = 2$) is already more accurate than Simpson's rule, we should expect a much better accuracy when we use $N = 3$ and higher.

Legendre polynomials are of course only a subset of orthogonal polynomials, any set of which could be used in the Gaussian integration formula. Although one can always transform any *finite* limits of integration to $(-1, 1)$, for integrals whose limits of integration include infinity, we shall have to use other orthogonal polynomials. Thus, if the interval of integration is $(-\infty, \infty)$, **Hermite polynomials** with the weight function $w(x) = e^{-x^2}$ are the "naturals." For $(0, \infty)$ one can use the **Laguerre polynomials** in which case the weight function is $w(x) = e^{-x}$. One may think that the appearance of the weight function would restrict the utility of Hermite and Laguerre polynomials. This is not the case, because one can always multiply and divide the integrand by the weight function. For example,

See *MM*, pp. 595–601 for Hermite and pp. 604–607 for Laguerre polynomials.

$$\int_0^\infty f(x)\,dx = \int_0^\infty e^{-x}\underbrace{[e^x f(x)]}_{\equiv g(x)}\,dx = \int_0^\infty e^{-x} g(x)\,dx$$

To use Gaussian integration formulas with different N, we need the actual values of the abscissas (evaluation points) as well as the weights W_i. Fortunately both of these are independent of the integrand: the abscissas are simply the roots of the orthogonal polynomials used, and the W_i are obtained from Equation (4.10). The values of abscissas and W_i for various N have been tabulated in books and handbooks on numerical analysis, so there is no need to calculate these from scratch.

4.3 Working with Series in *Mathematica*

We have thus far examined the infinite sequences and series of *numbers*. The real power of these discussions comes about when the terms of the sequence—and in particular series—are *functions* of a variable. There are a number of functions used as terms of an infinite series. One of them, $(x - x_0)^n$, and its special case in which $x_0 = 0$ are extremely useful.

Mathematica has the internal command `Series` for representing infinite series. For example,

```
In[1]:= Series[E^x,{x,0,5}]
```

which asks *Mathematica* to expand e^x about $x = 0$ up to order 5 in x, produces

$$1 + x + \frac{x^2}{2} + \frac{x^3}{6} + \frac{x^4}{24} + \frac{x^5}{120} + \mathrm{O[x]}^6$$

The symbol $\mathrm{O[x]}^6$, which in mathematics literature—*but not in Mathematica*—is also written as $O(x^6)$, signifies the fact that the next term is a multiple of x^6. This symbol is a signature of the expression `Series` and keeps track of the accuracy with which manipulations are to be performed. When O[x], raised to some power, appears in an expression, *Mathematica* treats that expression as an infinite series.

We can perform various operations on series. As as illustration, suppose we type in

```
In[1]:= S1:=Series[Sin[x],{x,0,5}]
```

with the output

$$x - \frac{x^3}{6} + \frac{x^5}{120} + \mathrm{O[x]}^6$$

and

```
In[2]:= S2:=Series[Log[1+x],{x,0,8}]
```

with the output

$$x - \frac{x^2}{2} + \frac{x^3}{3} - \frac{x^4}{4} + \frac{x^5}{5} - \frac{x^6}{6} + \frac{x^7}{7} - \frac{x^8}{8} + \mathrm{O[x]}^9$$

Then the command

```
In[3]:= S1+S2
```

produces

$$2x - \frac{x^2}{2} + \frac{x^3}{6} - \frac{x^4}{4} + \frac{5x^5}{24} + \mathrm{O[x]}^6$$

and all terms beyond power 5 are ignored. This is because we are ignorant of what happens to $S1$ beyond the fifth power, and this ignorance has to be carried over to the sum of the two series.

We can multiply two power series

```
In[4]:= S1 S2
```

and get

$$x^2 - \frac{x^3}{2} + \frac{x^4}{6} - \frac{x^5}{6} + \frac{11x^6}{72} + \mathrm{O[x]}^7$$

which may be surprising due to the appearance of $\mathrm{O[x]}^7$ and our ignorance of $S1$ beyond the fifth power. The explanation of this apparent contradiction is as follows. Think of $\mathrm{O[x]}^6$ as something like ax^6 where a is unknown. When the two series are multiplied, the term of lowest power that ax^6 can produce is ax^7 because the term of lowest power in $S2$ is x. If $S2$ had a lowest term of power different from 1, the result would be different. For example, type in

```
In[5]:= S3:=Series[Sqrt[1+x],{x,0,6}]
```

with the output

$$1 + \frac{x}{2} - \frac{x^2}{8} + \frac{x^3}{16} - \frac{5x^4}{128} + \frac{7x^5}{256} - \frac{21x^6}{1024} + \mathrm{O[x]}^7$$

and

```
In[6]:= S4:=Series[Sqrt[x^4+x^5],{x,0,6}]
```

with the output

$$x^2 + \frac{x^3}{2} - \frac{x^4}{8} + \frac{x^5}{16} - \frac{5x^6}{128} + \mathrm{O[x]}^7$$

Then the product `S1 S3` will yield

$$x + \frac{x^2}{2} - \frac{7x^3}{24} - \frac{x^4}{48} - \frac{19x^5}{1920} + \mathrm{O[x]}^6$$

and the product `S1 S4` will yield

$$x^3 + \frac{x^4}{2} - \frac{7x^5}{24} - \frac{x^6}{48} - \frac{19x^7}{1920} + \mathrm{O[x]}^8$$

When $\mathrm{O[x]}^6$ of $S1$ multiplies the lowest power of $S3$, it does not change, but when it multiplies the lowest power of $S4$, its power increases by 2.

Mathematica can divide two power series and get a third power series as the quotient. For example with $S1$ and $S2$ as above, the input `S1/S2` produces

$$1 + \frac{x}{2} - \frac{x^2}{4} - \frac{x^3}{24} - \frac{x^4}{240} + \mathrm{O[x]}^5$$

Once again, the power of O[x] has changed—it has decreased this time—because of the division by the lowest term of $S2$.

You can integrate a power series:

integrating a power series

```
In[1]:= Integrate[S1,x]
```

$$\frac{x}{2} - \frac{x^4}{24} + \frac{x^6}{720} + \mathrm{O[x]}^7$$

or differentiate a power series

```
In[1]:= D[S2,{x,3}]
```

$$2 - 6x + 12x^2 - 20x^3 + 30x^4 - 42x^5 + \mathrm{O[x]}^6$$

Note that integration raises the power of O[x] and differentiation lowers it.
Some commands pertaining to series are outlined below:

`Series[f[x],{x,a,n}]`	find the power series expansion of f about $x = a$ to order n
`Normal[series]`	convert a power series to a normal expression
`SeriesCoefficient[series,n]`	give the coefficient of the nth-order term in a power series
`LogicalExpand[series1==series2]`	give the equations obtained by equating coefficients of the two power series

4.4 Equations Involving Series

Many of the special functions of mathematical physics came into existence in the course of the eighteenth and nineteenth centuries as a result of the mathematicians' attempts at finding series solutions to (partial) differential equations (DEs). Later, when these series were studied further, they became separate entities worthy of careful and detailed analysis.

One of these functions is the set of **Legendre polynomials** $P_n(u)$. These are solutions of **Legendre DE** which arises when many of the partial DEs of mathematical physics are solved in spherical coordinates. Legendre polynomials have a **generating function** $g(t,u)$ given by

MM, Section 12.3.2

$$g(t,u) \equiv \frac{1}{\sqrt{1+t^2-2tu}}$$

As the name suggests, $g(t,u)$ *generates* the Legendre polynomials and does it by differentiation. More specifically, we have

$$\frac{1}{\sqrt{1+t^2-2tu}} = \sum_{n=0}^{\infty} t^n P_n(u) \tag{4.14}$$

which shows that $P_n(u)$ is, to within a constant, the nth derivative of $g(t,u)$ with respect to t evaluated at $t=0$. More precisely

MM, p. 547

$$P_n(u) = \frac{1}{n!}\frac{\partial^n}{\partial t^n}\left.\frac{1}{\sqrt{1+t^2-2tu}}\right|_{t=0} \tag{4.15}$$

We can write a simple *Mathematica* routine to calculate $P_n(u)$ for us. First we define the generating function

```
In[1]:= g[t_,u_]:=1/Sqrt[1+t^2-2 t u]
```

Next we evaluate its nth derivative with respect to t:

```
In[2]:= dn[t_,u_,n_]:=Evaluate[D[g[t,u],{t,n}]]
```

Finally we divide dn by $n!$ and evaluate the result at $t=0$ to obtain the Legendre polynomial of order n.

```
In[3]:= p[u_,n_]:=Simplify[dn[t,u,n]/n! /. t->0]
```

We can now make a table of the Legendre polynomials. For example, the command

```
In[4]:= Table[p[u,k],{k,0,4}]//TableForm
```

produces the following table:

$$\begin{array}{l} 1 \\ u \\ \frac{1}{2}(-1+3u^2) \\ \frac{1}{2}u(-3+5u^2) \\ \frac{1}{8}(3-30u^2+35u^4) \end{array}$$

which indeed lists the first five Legendre polynomials.

Our aim in this section is not, however, to practice differentiation or to generate Legendre polynomials,[3] but to acquaint ourselves with series manipulations. Instead of differentiating the generating function, we want to use its expansion to read off the Legendre polynomials.

Suppose we are interested in Legendre polynomials of order 5 or less. We could simply expand the generating function up to order 5 and read off the coefficients. So, we start with

```
In[1]:= ser=Series[1/Sqrt[1+t^2-2 t u],{t,0,5}]
```

and for any particular Legendre polynomial, we ask for the corresponding coefficient. For example,

SeriesCoefficient is used here.

[3]Incidentally, *Mathematica* has a built-in function called `LegendreP[n,x]`, which generates Legendre polynomials of any order.

```
In[2]:= SeriesCoefficient[ser,3]
```

produces

$$\frac{1}{3}\left(-2u+\frac{1}{2}u\left(-1+3u^2\right)\right)$$

which is the Legendre polynomial of order 3 in disguise and reveals its identity after applying `Simplify` to it.

This is fine, but it would be more convenient if we didn't have to type in `SeriesCoefficient[ser, ]` every time we needed a Legendre polynomial. We can achieve this by the command

```
In[3]:= legpol[k_,u_]:=Simplify[SeriesCoefficient[ser,k]]
```

Then, `legpol[k,u]` produces the Legendre polynomial of order k *as long as* k *is less than* 5, the highest power of the expansion of `ser`.

Such a fixation of the order of `ser` is inconvenient, because for higher-order polynomials, we will have to change the order and start from scratch. Thus, a further improvement would be to replace 5 with a variable. So, let us define

```
In[4]:= LegSer[t_,u_,m_] := Series[1/Sqrt[1+t^2-2t u],
        {t, 0, m}]
```

Although *Mathematica* complains about m not being a machine-size integer, the scheme will work. All we need to do now is define the polynomials we want:

```
In[5]:= LPol[k_,u_]:=Simplify[SeriesCoefficient
        [LegSer[t,u,m],k]]
```

Then, `LPol[k,u]` produces the Legendre polynomial of order k *as long as* k *is less than* m. To ensure this, we simply choose m to be $k+1$. Thus, finally, we obtain

```
In[6]:= LegPol[k_,u_]:=Simplify[SeriesCoefficient
        [LegSer[t,u,k+1],k]]
```

where `LegSer` is defined in `In[4]`. Now we can generate Legendre polynomials of any order.

A third way of generating the Legendre polynomials is to create two series, one for the left-hand side of (4.14) and one for its right-hand side, equate the corresponding coefficients, and solve the resulting equations. The reason that we even discuss this (clumsy) procedure is to get acquainted with the steps involved.

Define a series for the generating function

```
In[1]:= GenFnSer=Series[1/Sqrt[1+t^2-2 t u],{t,0,3}];
```

and the series for the right-hand side of (4.14)

```
In[2]:= rhsSer=Sum[p[n] t^n,{n,0,3}]+O[t]^4;
```

The term `O[t]^4` turns the sum into a series. Now equate the corresponding coefficients of the two series:

```
In[3]:= equations=LogicalExpand[GenFnSer==rhsSer]
```

and get

$$1 - p[0] == 0 \ \&\& \ u - p[1] == 0 \ \&\& \ \frac{1}{2}(-1 + 3u^2) - p[2] == 0 \ \&\&$$
$$\frac{1}{3}\left(-2u + \left(\frac{5}{2}u(-1 + 3u^2)\right)\right) - p[3] == 0$$

Now solve these equations

```
In[4]:= solution=Solve[equations,{p[0],p[1],p[2],p[3]}]
```

to get

$$\left\{\left\{p[0] \to 1, p[1] \to u, p[2] \to \frac{1}{2}(-1 + 3u^2), p[3] \to \frac{1}{2}(-3u + 5u^3),\right\}\right\}$$

But we want the polynomials, not the assignment rules. A code that does this is

```
In[5]:= Do[p[n]=First[p[n]/.solution],{n,0,3}]
```

where `p[n]/.solution` picks $p[n]$ with curly brackets still around it, and `First` picks the first—in this case the only—element of what is in the curly brackets. Now the Legendre polynomials are available for use. Typing

```
In[6]:= Table[p[n],{n,0,3}]
```

produces

$$\left\{1, u, \frac{1}{2}(-1 + 3u^2), \frac{1}{2}(-3u + 5u^3)\right\}$$

The technique just outlined is used to find the power series representation of the solutions of differential equations. We solve the **Hermite differential equation**

$$H'' - 2xH' + 2nH = 0 \tag{4.16}$$

and construct **Hermite polynomials**.

MM, pp. 593–601, discusses Hermite polynomials.

We start with the unknown series

```
In[1]:= y=Sum[a[k] x^k,{k,0,8}]+O[x]^9;
```

Note that `O[x]^9` is essential as it makes the above expression a *series*, not just a sum. The idea is to find `a[k]` in such a way that `y` satisfies Equation (4.16). Thus, we substitute y in the left-hand side of that equation

```
In[2]:= lhs=Evaluate[D[y,{x,2}]]-2x Evaluate[D[y,x]]
        +2n y;
```

and set it equal to zero and tell *Mathematica* to find the equations that determine the coefficients `a[k]`:

```
In[3]:= equations=LogicalExpand[lhs==0]
```

The output will look like

$$2na[0]+2a[2]==0\ \&\&\ -2a[1]+2na[1]+6a[3]==0\&\&$$
$$-4a[2]+2na[2]+12a[4]==0\ \&\&\ -6a[3]+2na[3]+20a[5]==0\&\&$$
$$-8a[4]+2na[4]+30a[6]==0\ \&\&\ -10a[5]+2na[5]+42a[7]==0\&\&$$
$$-12a[6]+2na[6]+56a[8]==0$$

It turns out that if n in (4.16) is an integer (as it usually is in physical applications), the solution will be a polynomial of degree n. Furthermore, if n is even (odd), only even (odd) powers of x will appear in the polynomial, and the coefficients of even (odd) powers of x will be a multiple of `a[0]` (`a[1]`). Therefore, in solving the equations above for coefficients, we will ignore `a[0]` and `a[1]`. So, the next step is to type in

```
In[4]:= answers=Simplify[Solve[equations,{a[2],a[3],
        a[4],a[5],a[6],a[7],a[8]}]]
```

and get the output

$$\Big\{\Big\{a[7]\to-\frac{1}{630}(-5+n)(-3+n)(-1+n)a[1],$$
$$a[8]\to\frac{(-6+n)(-4+n)(-2+n)na[0]}{2520},$$
$$a[5]\to\frac{1}{30}(-3+n)(-1+n)a[1],$$
$$a[6]\to-\frac{1}{90}(-4+n)(-2+n)na[0],a[4]\to\frac{1}{6}(-2+n)na[0],$$
$$a[2]\to-na[0],a[3]\to-\frac{1}{3}(-1+n)a[1]\Big\}\Big\}$$

Notice how all `a[k]` with even k are multiples of `a[0]`, and those with odd k are multiples of `a[1]`. Note also that if n is even (odd), the even (odd) coefficients will be zero after a certain point. For example, if $n=4$, `a[6]` and `a[8]` will be zero.[4] Just substitute 4 for n in the expressions for `a[6]` and `a[8]`, and note that they vanish.

Having found the coefficients, we substitute them in y. This is done as follows:

[4] `a[8]` is the highest coefficient in our expansion. If there were higher coefficients, they would be zero as well.

```
In[5]:= y=y/.answers
```

This yields

$$\Big\{a[0]+a[1]x-na[0]x^2$$
$$-\frac{1}{3}(-1+n)a[1]x^3+\frac{1}{6}(-2+n)na[0]x^4$$
$$+\frac{1}{30}(-3+n)(-1+n)a[1]x^5-\frac{1}{90}(-4+n)(-2+n)na[0]x^6$$
$$-\frac{1}{630}(-5+n)(-3+n)(-1+n)a[1]x^7$$
$$+\frac{(-6+n)(-4+n)(-2+n)na[0]}{2520}x^8+O[x]^9\Big\}$$

Since we are interested in the expression in the curly brackets, we get rid of them by typing

```
In[6]:= First[%]
```

Next, we get rid of $O[x]^9$ by typing

```
In[6]:= h=Normal[%]
```

We now have everything at our disposal to create a table of Hermite polynomials for varying n. If we repeat h for different n, we obtain these polynomials to within a factor that depends on the order of the polynomial. These factors have been determined by convention, and even without them h (or y) will satisfy the Hermite differential equation.[5] Thus, *for even polynomials* we multiply h by $2(-1)^{(n/2)}(n-1)!/(n/2-1)!$, and type in

```
In[7]:= tab=Table[Simplify[2h(-1)^(n/2)(n-1)!/
        ((n/2-1)!)]/.{a[0]->1,a[1]->0},{n,2,8,2}]
```

to produce the following table:

$$\Big\{2(-1+2x^2),4(3-12x^2+4x^4),8(-15+90x^2-60x^4+8x^6),$$
$$1680\left(1-8x^2+8x^4-\frac{32x^6}{15}+\frac{16x^8}{105}\right)\Big\}$$

Note that we set $a[1]$ equal to zero to eliminate all odd powers of x. The table shows the even Hermite polynomials of order 2 through 8. For higher-order polynomials, we raise the power of x in the series y at the beginning. If we are interested in odd polynomials, we have to multiply h by

$$\frac{2(-1)^{(n-1)/2}n!}{((n-1)/2)!}$$

[5]If we multiply a solution of Equation (4.16) by a constant factor, it will still satisfy the equation.

set $a[0]$ equal to zero and $a[1]$ equal to 1. In fact,

```
In[8]:= tab=Table[Simplify[2h(-1)^((n-1)/2)n!/
        (((n-1)/2)!)]/.{a[0]->0,a[1]->1},{n,3,7,2}]
```

produces

$$\left\{-12\left(x-\frac{2x^3}{3}\right), 8x(15-20x^2+4x^4),\right.$$
$$\left.-1680\left(x-2x^3+\frac{4x^5}{5}-\frac{8x^7}{105}\right)\right\}$$

4.5 Fourier Series

Power series are extremely useful in representing functions. In fact, many (if not most of the) functions of mathematical physics were discovered first in the form of a power series. In many physical and engineering applications, however, series that use functions other than powers of x (or $x - x_0$) tend to be more convenient. **Fourier series** use the trigonometric functions (sines and cosines) as the basis of expansion. Because of their oscillatory nature, expansion in terms of the trigonometric functions results in functions that are *periodic*. This is particularly useful in electrical engineering, where periodic signals of various shapes are used extensively.

MM, pp. 243–247

Suppose we have a periodic function $f(t)$, defined in the interval (a, b) whose period is T, so that $f(t+T) = f(t)$. Then it can be shown that $f(t)$ can be written as an infinite series of sine and cosine terms as follows:

$$f(t) = a_0 + \sum_{n=1}^{\infty}\left(a_n \cos\frac{2n\pi t}{T} + b_n \sin\frac{2n\pi t}{T}\right) \tag{4.17}$$

where

$$\begin{aligned} a_0 &= \frac{1}{T}\int_a^b f(t)\,dt \\ a_n &= \frac{2}{T}\int_a^b f(t)\cos\frac{2n\pi t}{T}\,dt \\ b_n &= \frac{2}{T}\int_a^b f(t)\sin\frac{2n\pi t}{T}\,dt \end{aligned} \tag{4.18}$$

Thus, if we calculate a_0, a_n, and b_n, and insert them in Equation (4.17), in principle, we obtain a series representation of $f(t)$. In practice, we keep a finite number of terms of the series and obtain an *approximation* of the function in terms of a finite sum. Applied to the production of periodic

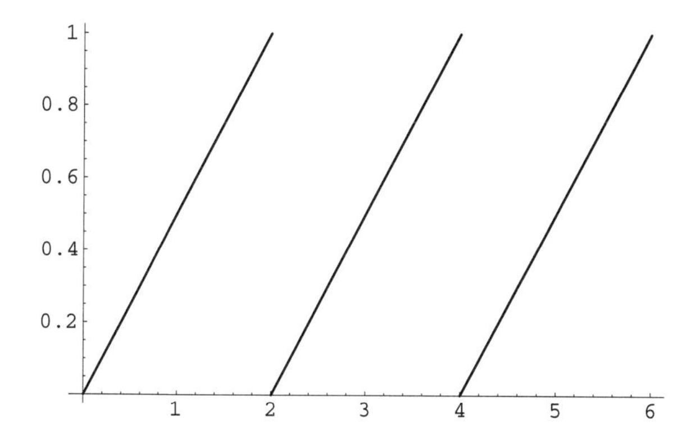

FIGURE 4.4. This ideal sawtooth potential can be approximated by a finite sum of sines and cosines of appropriate amplitude (coefficient) and frequency.

voltages of various shapes such as the one shown in Figure 4.4, this corresponds to adding a finite number of sine and cosine terms with appropriate coefficients and frequencies.

As Figure 4.4 indicates, a Fourier series is often used to represent discontinuous functions or functions that have different "pieces" in different regions of their interval of definition. *Mathematica* is capable of defining such discontinuous functions. It does so using logical operators as discussed in Section 2.6. In the following we make use of this technique to construct functions, and use that construction to find their Fourier series.

To construct the Fourier series of a function, we need the coefficients of the sine and cosine terms for the function at hand. We want to be as general as we can, so we type in

```
In[1]:= aNaught[T_,t0_]:=(1/T) Integrate[f[t],{t,t0,t0+T}];
        anFS[n_,T_,t0_]:=(2/T)Integrate[f[t]
        Cos[2 Pi n t/T],{t,t0,t0+T}];
        bnFS[n_,T_,t0_]:=(2/T)Integrate[f[t]
        Sin[2 Pi n t/T],{t,t0,t0+T}]
```

where $(t_0, t_0 + T)$ is the interval in which the function is defined and T is the period of the function. These formulas can be used for any function $f(t)$, which we have to define separately. The best way to explain all of this is to go through some examples in detail.

Let us find the Fourier series for the periodic voltage known as the **square wave**. This is shown in Figure 4.5, where the voltage is seen to oscillate between 1 volt (lasting for 2 seconds) and 0 (lasting for 1 second), so that the period of the function is 3 seconds. The function that describes this voltage in the interval $(0, 3)$ is

```
In[2]:= f[t_]:=If[t<2,1,0]
```

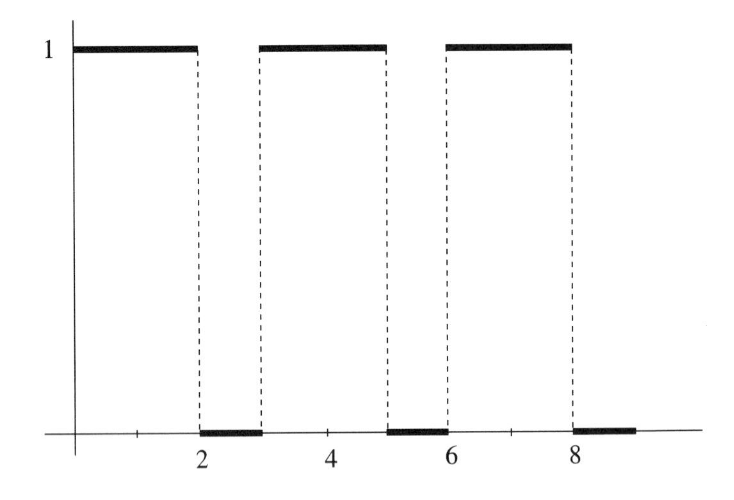

FIGURE 4.5. The square wave is a periodic voltage that oscillates between two constant values. In this graph, the two values are 1 volt (lasting for 2 seconds) and 0 (lasting for 1 second).

To find the coefficients of the Fourier series of this function, we execute the integrals of `In[1]` above, and then type in

```
In[3]:= a0=aNaught[3,0]; a[n_]:=anFS[n,3,0];
        b[n_]:=bnFS[n,3,0]
```

As a check, we tell *Mathematica* to output a_n,

```
In[4]:= a[n]
```

$$\frac{\sin\left(\frac{4n\pi}{3}\right)}{n\pi}$$

and b_n

```
In[5]:= b[n]
```

$$-\frac{-1+\cos\left(\frac{4n\pi}{3}\right)}{n\pi}$$

which are the results we expect.

Now that we have the coefficients, we can write down the Fourier series *as a finite sum* with a variable number of terms:

```
In[6]:= FourierSer[m_,t_,T_]:=a0+Sum[a[k]Cos[2k Pi t/T]
        +b[k]Sin[2k Pi t/T],{k,1,m}]
```

And if we are interested in seeing what `FourierSer[m,t,T]` looks like for various values of m, we can plot it. For example,

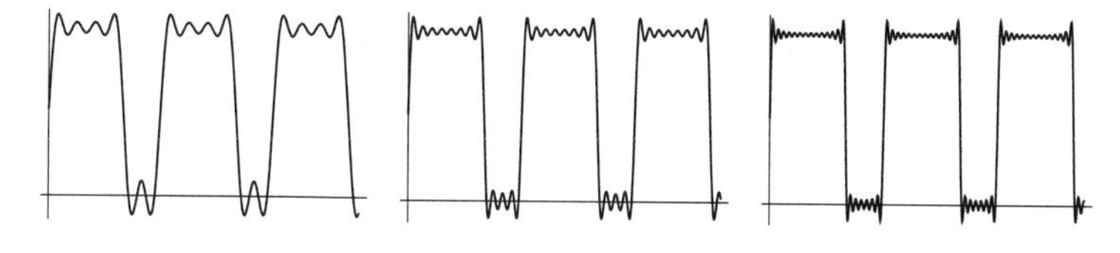

FIGURE 4.6. The finite-sum approximation to the Fourier series representation of the square wave voltage with 5 terms (left), 10 terms (middle), and 20 terms (right).

```
In[7]:= Plot[Evaluate[FourierSer[5,t,3]],{t,0,8.5}]
```

produces the diagram on the left in Figure 4.6. The command `Evaluate` causes a much faster calculation for the plot. Changing 5 to 10 and 20 will produce the middle and the right diagrams, respectively.

The **sawtooth wave** shown in Figure 4.4 is a common voltage in electrical engineering, and we can easily construct its Fourier series representation. First we define the function in the interval $(0, 2)$. It is a function that rises linearly from 0 to 1 volt in 2 seconds, which is its period T. So, $f(t) = t/2$, and we type in

```
In[8]:= f[t_]:=t/2
```

and execute the integrals defined earlier in `In[1]`. Next, we define the coefficients

```
In[9]:= a0=aNaught[2,0]; a[n_]:=anFS[n_,2,0];
        b[n_]:=bnFS[n_,2,0]
```

As a check, we note that

```
In[10]:= a[n]
```

yields

$$\frac{-1 + \cos(2n\pi) + 2n\pi \sin(2n\pi)}{2n^2\pi^2}$$

and

```
In[11]:= b[n]
```

yields

$$\frac{-2n\pi \cos(2n\pi) + \sin(2n\pi)}{2n^2\pi^2}$$

which simplify to $a_n = 0$ and $b_n = -1/(2n\pi)$, respectively, but since *Mathematica* does not know that n is an integer, it will not simplify the expressions any further.

Now we can execute `FourierSer[m,t,T]` of `In[6]` above to obtain the finite-sum approximation to the Fourier series expansion of the sawtooth wave. To see what `FourierSer[m,t,T]` looks like, we plot it for various values of m. For example,

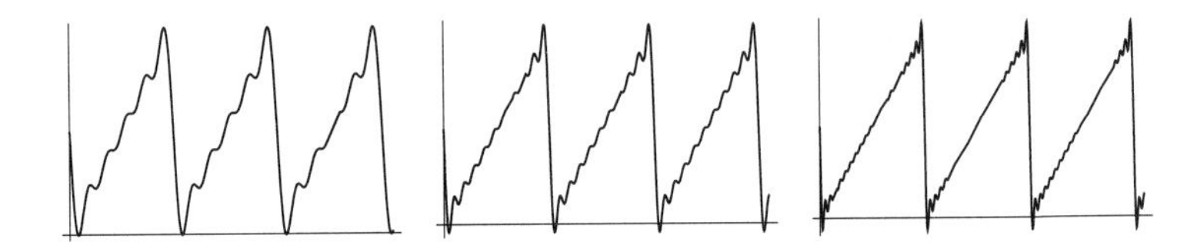

FIGURE 4.7. The finite-sum approximation to the Fourier series representation of the sawtooth voltage with 5 terms (left), 10 terms (middle), and 20 terms (right).

```
In[12]:= Plot[Evaluate[FourierSer[5,t,2]],{t,0,6.2}]
```

produces the diagram on the left in Figure 4.7. Changing 5 to 10 and 20 will produce the middle and the right diagrams, respectively.

4.6 Problems

Problem 4.1. Write a single integral giving the area of a unit circle. Approximate the integral with a sum of 10 terms corresponding to each of the methods of numerical integration discussed in Section 4.2. Which one is the best approximation?

Problem 4.2. The *gamma function* is defined as

$$\Gamma(x) = \int_0^\infty t^{x-1} e^{-t}\, dt$$

Choose a large value for the upper limit of integration and evaluate the integral using a finite sum corresponding to each of the methods of Section 4.2 to calculate $\Gamma(3/2)$. Compare your results with the "exact" value produced by *Mathematica*. (*Mathematica* has an internal gamma function.)

Problem 4.3. The *complete elliptic integrals* of the first and second kinds are defined as

$$K(x) = \int_0^{\pi/2} \frac{dt}{\sqrt{1 - x^2 \sin^2 t}} \quad \text{and} \quad E(x) = \int_0^{\pi/2} \sqrt{1 - x^2 \sin^2 t}\, dt$$

respectively. Evaluate each of these integrals using a finite sum corresponding to each of the methods of Section 4.2 to calculate $K(1/2)$ and $E(1/2)$. Compare your results with the "exact" values produced by *Mathematica.*

Problem 4.4. The molar heat capacity of a solid as a function of temperature is given by [Call 85, p. 367]

$$c(\tau) = 9R\tau^3 \int_0^{1/\tau} \frac{x^4 e^x}{(e^x - 1)^2}\, dx$$

where R is the universal gas constant and τ is the temperature measured in units of the *Debye temperature.*
(a) For each of the different methods of integration of Section 4.2, define $c(\tau, n)$ to be the approximation to the above integral in terms of a sum of n terms.
(b) Plot $c(\tau, n)/R$ as a function of τ for large enough n for each method. Choose the range of τ to be from 0.01 to 2.
(c) Using `NIntegrate` in *Mathematica*, plot $c(\tau)$ for the same range of τ as in (b) and compare the graph with the graphs obtained in (b).

Problem 4.5. Hermite and Laguerre polynomials of second order are

$$4x^2 - 2 \quad \text{and} \quad \tfrac{1}{2}(x^2 - 4x + 2)$$

respectively. Use Gauss–Hermite or Gauss–Laguerre integration method (similar to Gauss–Legendre method of Section 4.2.4) to find the numerical values of the following integrals.

$$\text{(a)} \int_{-\infty}^{\infty} \frac{dx}{1+x^4} \qquad \text{(b)} \int_{0}^{\infty} \frac{dx}{1+x^3} \qquad \text{(c)} \int_{-\infty}^{\infty} \frac{dx}{1+x^6}$$

$$\text{(d)} \int_{0}^{\infty} \frac{e^{-x^4}}{1+x^2} dx \quad \text{(e)} \int_{0}^{\infty} \frac{e^{-x}}{1+x^3}\, dx \quad \text{(f)} \int_{-\infty}^{\infty} \frac{e^{-x^2}}{1+x^6}\, dx$$

Problem 4.6. (a) See if *Mathematica* can sum the following series in closed form:

$$\sum_{n=1}^{\infty} \frac{\sin n}{n^2}, \quad \sum_{n=2}^{\infty} \frac{1}{n^3 - 1}, \quad \sum_{n=1}^{\infty} \frac{n+5}{n^2 - 3n - 5}, \quad \sum_{n=2}^{\infty} \frac{1}{\sqrt{n}\,\ln n}$$

$$\sum_{n=1}^{\infty} \frac{1}{n^2+1}, \quad \sum_{n=1}^{\infty} \frac{n}{n^2+1}, \quad \sum_{n=2}^{\infty} \frac{1}{n\,\ln^2 n}, \quad \sum_{n=2}^{\infty} \frac{1}{n\,\ln n \ln \ln n}$$

$$\sum_{n=1}^{\infty} \frac{2^n + 1}{3^n + n}, \quad \sum_{n=1}^{\infty} \frac{(-1)^n}{n!}, \quad \sum_{n=1}^{\infty} \frac{5^n}{n!}$$

(b) By summing 20, 50, and 100 terms of each series, decide if the series converges (or diverges).

Problem 4.7. Find the power series expansion to order 7 of the following functions about $x = 0$.

$$\text{(a)}\ \tan^{-1} x \qquad \text{(b)}\ \frac{\sin x}{e^x} \qquad \text{(c)}\ \tanh x$$

$$\text{(d)}\ \frac{\ln(1+x)}{\cos x} \qquad \text{(e)}\ \frac{\sqrt{1+x}}{e^x \cos x} \qquad \text{(f)}\ \tan x \tanh x$$

$$\text{(g)}\ \frac{\ln(1+x^2)}{\sqrt{4+x^2}} \qquad \text{(h)}\ \frac{\sqrt{1+\cos x}}{e^{x^2}} \qquad \text{(i)}\ \tanh(\sin x)$$

Problem 4.8. Find the power series expansion to order 8 of the following products of functions about $x = 0$ in two ways: directly, and by multiplying the series of each factor in the product.

(a) $\tan^{-1} x \sin x$ (b) $\cos x e^x$ (c) $\tanh x \ln(3 + 2x)$

(d) $\dfrac{\ln(1 + x^2)}{\cos x}$ (e) $\sqrt[4]{1 + x}e^{-x}$ (f) $\sin x \sinh x$

Problem 4.9. Using the technique of coefficient matching as done for Hermite polynomials on page 144, find the power series solution of the following differential equations:

(a) $y' + y = 0$ (b) $y'' - y = 0$ (c) $y'' + y = 0$
(d) $y'' + y' + y = 0$ (e) $y'' + 2y' + y = 0$ (f) $y'' + 5y' + 4y = 0$
(g) $y' + 2xy = 0$ (h) $y' + 3x^2 y = 0$

Compare your series with the series of the known solutions obtained using *Mathematica.*

Problem 4.10. A voltage $V(t)$ is given by

$$V(t) = \begin{cases} 2t & \text{if } 0 \le t \le 1 \\ 0 & \text{if } 1 \le t \le 2 \end{cases}$$

If this voltage repeats itself periodically, find the Fourier series expansion of $V(t)$ and plot the series for 3, 5, and 30 terms and $0 \le t \le 6$.

Problem 4.11. A periodic voltage with period 2 is given by

$$V(t) = \begin{cases} \cos(\pi t) & \text{if } -1/2 \le t \le 1/2 \\ 0 & \text{if } 1/2 \le |t| \le 1 \end{cases}$$

(a) Find the Fourier series of $V(t)$.
(b) Evaluate both sides at $t = 0$ to show that

$$\frac{\pi}{2} = 1 - 2\sum_{n=1}^{\infty} \frac{(-1)^n}{4n^2 - 1}$$

This is one of the many series representations of π.
(c) How many terms of the series do you have to keep for the right-hand side to agree with the left-hand side to six decimal places?

Problem 4.12. An electric voltage $V(t)$ is given by

$$V(t) = \sin\left(\frac{\pi t}{2}\right)$$

and repeats itself with period 1. Find the Fourier series expansion of $V(t)$, and plot it for 3, 5, and 30 terms and $0 \le t \le 4$.

Problem 4.13. A periodic voltage is given by the formula

$$V(t) = \begin{cases} \sin(\pi t/2) & \text{if } \ 0 \leq t \leq 1 \\ 0 & \text{if } \ 1 \leq t \leq 2 \end{cases}$$

Find the Fourier series representation of this voltage, and plot it for 3, 5, and 30 terms and $0 \leq t \leq 6$.

Problem 4.14. A periodic voltage with period 4 is given by

$$V(t) = \begin{cases} (1 - t^2) & \text{if } \ |t| \leq 1 \\ 0 & \text{if } \ 1 \leq |t| \leq 2 \end{cases}$$

Find the Fourier series of $V(t)$, and plot it for 3, 5, and 30 terms and $-6 \leq t \leq 6$.

5

Numerical Solutions of ODEs: Theory

The ubiquity of differential equations in all areas of physics and their resistance to analytical solutions in many cases of interest, plus the availability of cheaper and cheaper high-speed computers, has led to an explosion of interest in the numerical solution of differential equations. The *techniques* used in solving DEs are not new; some of the earlier ones date back to Euler himself.

Although unnecessary for using *Mathematica* to solve differential equations, the study of the techniques of the numerical solutions of DEs is important enough that it is worth our time and effort understanding their essential elements. Therefore, this chapter has very little to do with *Mathematica*, but a lot to do with what is happening "behind the scene," so to speak.

We start with a very general first-order DE written in the form

$$y' = f(x, y) \qquad \text{or} \qquad \frac{dy}{dx} = f(x, y) \tag{5.1}$$

The solution of this equation is a function $y(x)$ whose derivative is equal to $f(x, y(x))$. Let us assume that we know the solution at some *initial* value of x, say x_0, so that

$$y'(x_0) = f(x_0, y(x_0)) = f(x_0, y_0) \quad \text{where} \quad y_0 \equiv y(x_0) \tag{5.2}$$

Any procedure for solving a DE will have to start with the initial condition (x_0, y_0) and generate a (large) set of pairs (x_k, y_k) interpreted as a tabular representation of the function $y(x)$. This requires being able to find $y(x_0+h)$

from the initial condition (5.2) and the DE (5.1). Here and in the sequel, h is a (small) step size.

Many of the common procedures can be obtained by "integrating" (5.1) from x_0 to $x_0 + h$:

$$y(x_0 + h) = y(x_0) + \int_{x_0}^{x_0+h} f(t, y(t))\, dt \tag{5.3}$$

This is not, of course, a solution of the problem because the integrand contains $y(t)$, which is the unknown function we are after! However, since we know of different ways of approximating the integral, we may have a chance of finding an *approximate* solution to the DE.

This approximation starts with noting that since h is small, we can assume that the function f is continuous in the interval of integration, allowing us to invoke the mean value theorem of calculus and write[1]

$$y(x_0 + h) = y(x_0) + hf(c, y(c)) \equiv y(x_0) + hf(x_0 + \alpha h, y(x_0 + \alpha h)) \tag{5.4}$$

where $x_0 \leq c \leq x_0 + h$ and $0 \leq \alpha \leq 1$. Equation (5.4) is *exact*, and therefore, not a solution to our DE. Although the mean value theorem proves the *existence* of c (or α), it gives no clue as to how to find its actual value. Thus, we are led to approximating c or α.

5.1 Various Euler Methods

The accuracy of the numerical solution of (5.1) depends on the sophistication with which we approximate the integral of (5.3). Section 4.2 treated a number of approximations to an integral. We shall employ those approximations to obtain solutions to our DE with varying degree of accuracy.

5.1.1 *Euler Method*

If we let α assume its lowest possible value, namely zero, we obtain

$$y_1 \equiv y(x_0 + h) = y(x_0) + hf(x_0, y(x_0)) = y(x_0) + hf(x_0, y_0) \equiv y_0 + hf_0 \tag{5.5}$$

where we have introduced the notation y_1 for $y(x_0+h)$ and f_0 for $f(x_0, y_0)$. We shall also use x_k for $x_0 + kh$. Iterating (5.5) yields

$$y_2 \equiv y(x_0+2h) = y(x_0+h)+hf(x_0+h, y(x_0+h)) = y_1+hf(x_1, y_1) \equiv y_1+hf_1$$

[1] Recall that the mean value theorem says that if $g(t)$ is a continuous function on (a, b), then $\int_a^b g(t)\, dt = (b - a)g(c)$ for some c in the interval (a, b).

and, in general,

$$y_{k+1} = y_k + hf_k, \qquad k = 0, 1, \ldots \tag{5.6}$$

Thus, starting with x_0, y_0, and f_0, we can generate as long a table as we desire representing the solution of the DE (5.1). The problem with this **Euler method** is that it is very crude, and unless h is (unreasonably) small, the method will not give a good approximation to the solution.

Euler method

5.1.2 *Modified Euler Method*

An improvement over the Euler method is obtained by taking c to be the midpoint of the integration interval. This corresponds to $\alpha = 1/2$ and

$$y(x_0 + h) = y(x_0) + hf(x_0 + h/2, y(x_0 + h/2)) \tag{5.7}$$

The problem is that we do not know what $y(x_0 + h/2)$ is! However, we can *approximate* it. How? By Taylor series:

$$y(x_0 + h/2) \approx y(x_0) + \frac{h}{2}y'(x_0) = y(x_0) + \frac{h}{2}f(x_0, y_0) = y_0 + \frac{h}{2}f_0 \tag{5.8}$$

It follows that

$$y_1 \equiv y(x_0 + h) = y(x_0) + hf(x_0 + h/2, y_0 + h/2 f_0)$$

and, in general,

$$y_{k+1} = y_k + hf(x_k + h/2, y_k + (h/2)f_k), \qquad k = 0, 1, \ldots \tag{5.9}$$

where $f_k = f(x_k, y_k) = f(x_0 + kh, y(x_0 + kh))$. Equation (5.9) is called the **modified Euler method**.

modified Euler method

Figure 5.1 shows the difference between Euler and modified Euler approximations to the integral. Generally speaking, by taking the height of the elementary rectangles to be the value of the function evaluated at the midpoint of the interval $(x_0, x_0 + h)$, we cover parts of the actual area that are missed in the simple Euler method.

5.1.3 *Improved Euler Method*

In our discussion of numerical integration in Section 4.2, we came across techniques that gave more accurate results than the sum of simple rectangles. Can we use those techniques to improve our numerical solution of ODEs? Let us try the trapezoid rule first. Using Equation (4.4) we obtain

$$\begin{aligned} y(x_0 + h) &= y(x_0) + \int_{x_0}^{x_0+h} f(t, y(t))\, dt \\ &= y(x_0) + \frac{h}{2}[f_0 + f(x_0 + h, y(x_0 + h))] \end{aligned} \tag{5.10}$$

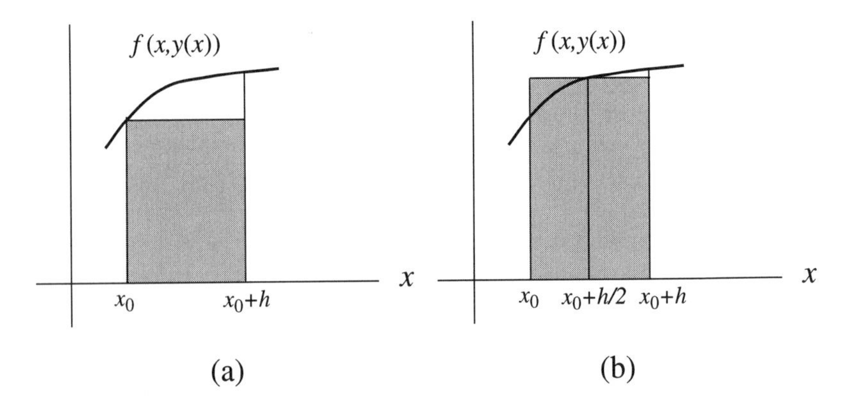

FIGURE 5.1. (a) The Euler method corresponds to approximating the area by the rectangle whose height is the value of the function at the initial point. (b) The modified Euler method corresponds to approximating the area by the rectangle whose height is the value of the function at the midpoint. We expect the latter to be a better approximation to the actual area.

Now use Taylor series expansion to approximate $y(x_0+h)$ *on the right-hand side*:

$$y(x_0 + h) \approx y(x_0) + hy'(x_0) = y_0 + hf_0$$

It now follows that

$$y_1 \equiv y(x_0 + h) \approx y_0 + \frac{h}{2}[f_0 + f(x_0 + h, y_0 + hf_0)]$$

improved Euler method

and, in general,

$$y_{k+1} \approx y_k + \frac{h}{2}[f_k + f(x_k + h, y_k + hf_k)] \tag{5.11}$$

This is called the **improved Euler method**.

5.1.4 *Euler Methods in* Mathematica

We can write a *Mathematica* routine to solve a first-order DE numerically using any one of the Euler methods discussed so far. The idea is to make a table with the first column holding equipartitioned values of the independent variable x and the second column holding the corresponding values of the dependent variable y.

The first line of this routine initializes the independent and dependent variables to x_0 and y_0:

```
In[1]:= x[0]=x0; yEu[0]=y0; yMEu[0]=y0; yIEu[0]=y0;
```

with the obvious notation that yEu, yMEu, and yIEu represent the dependent variable obtained using the Euler, modified Euler, and improved Euler methods, respectively.

Next, we create the values of the independent variable using the Do command:

```
In[2]:= IndVar[h_, n_] := Do[x[i] = x0 + i h, {i, n}];
```

This creates an array of $n+1$ values x_i by adding multiples of h to the initial value x_0. The values of the independent variable so created are now ready to be used in the DE to obtain the corresponding values of the dependent variable.

To create these values, we use the three Euler methods. For the simplest Euler method, we employ Equation (5.6), and type in

```
In[3]:= Eul[h_,n_]:=Do[yEu[i+1]=yEu[i]+h f[x[i],yEu[i]],
        {i, 0, n}];
```

For the modified Euler method, we use Equation (5.9):

```
In[4]:= MEul[h_,n_]:=Do[yMEu[i+1]=yMEu[i]+h f[x[i]+h/2,
        yMEu[i]+(h/2) f[x[i],yMEu[i]]],{i, 0, n}];
```

Finally, for the improved Euler method, we make use of Equation (5.11):

```
In[5]:= IEul[h_,n_]:=Do[yIEu[i+1]=yIEu[i]+(h/2)(f[x[i],
        yIEu[i]]+f[x[i]+h,yIEu[i]+h f[x[i], yIEu[i]]]),
        {i, 0, n}];
```

In all the above statements, we have suppressed the output by using semicolons (although the use of := postpones the evaluation anyway).

As a special example, let us consider the simple DE

$$y' = y, \qquad x_0 = 0, \quad y_0 = 1$$

whose solution is known to be e^x. To evaluate various solutions, we input the initial data and the function for this particular DE:

```
In[6]:= x0 = 0; y0 = 1; f[x_, y_] := y;
```

Then evaluate the array of the independent variable for $h = 0.05$ and $n = 20$:

```
In[7]:= IndVar[0.05, 20];
```

Finally, the input

```
In[8]:= Eul[0.05, 20]; MEul[0.05, 20]; IEul[0.05, 20];
```

creates the array of the independent variable using the three Euler methods. To display these functions in a table, we type in

```
In[9]:= Table[{x[i],yEu[i],yMEu[i],yIEu[i], Exp[x[i]]},
        {i,0,20,2}]// MatrixForm
```

x	Euler	Modified Euler	Improved Euler	e^x
0.0	1	1	1	1
0.1	1.10250	1.10513	1.10513	1.10517
0.2	1.21551	1.22130	1.22130	1.22140
0.3	1.34010	1.34970	1.34970	1.34986
0.4	1.47746	1.49159	1.49159	1.49182
0.5	1.62889	1.64839	1.64839	1.64872
0.6	1.79586	1.82168	1.82168	1.82212
0.7	1.97993	2.01319	2.01319	2.01375
0.8	2.18287	2.22483	2.22483	2.22554
0.9	2.40662	2.45871	2.45871	2.45960
1.0	2.65330	2.71719	2.71719	2.71828

TABLE 5.1. Comparison of the Euler, modified Euler, and improved Euler formulas with the exact solution for $f(x,y) = y$. It is clear that modified and improved Euler formulas give a more accurate result than the simple Euler method. The equality of the modified and improved Euler methods is a coincidence.

and obtain a matrix with values as given in Table 5.1. Notice that although the step size is 0.05 for better accuracy, the displayed step size is 0.1, because we have specified the range of the index i by $\{i, 0, 20, 2\}$, in which the last number gives the increment of i. The table shows that the modified Euler and improved Euler methods give a much closer result to the exact answer than the simple Euler method. The fact that modified Euler and improved Euler results are the same is a coincidence brought about by the special nature of $f(x,y)$ in this example.

To differentiate between the last two Euler methods, let us look at another differential equation. For $f(x,y)$, take the function $-2xy$, which again renders the DE analytically soluble, with the solution e^{-x^2} if we impose the same initial data as before. So, we type in

```
In[10]:= x0 = 0; y0 = 1; f[x_, y_] := -2 x y;
```

and

```
In[11]:= Eul[0.05, 20]; MEul[0.05, 20]; IEul[0.05, 20];
```

to reevaluate the functions for the new $f(x,y)$. The command

```
In[12]:= Table[{x[i],yEu[i],yMEu[i],yIEu[i],
         Exp[-x[i]^2]},{i,0,20,2}]// MatrixForm
```

then produces a matrix whose entries are shown in Table 5.2. The table makes it clear that the simple Euler method is a poor approximation, while the other two Euler methods give more accurate results, with the improved Euler method gaining a little advantage over the modified method.

x	Euler	Modified Euler	Improved Euler	e^{-x^2}
0.0	1	1	1	1
0.1	0.995	0.990037	0.990044	0.990050
0.2	0.970274	0.960742	0.960779	0.960789
0.3	0.927097	0.913833	0.913920	0.913931
0.4	0.867809	0.851987	0.852141	0.852144
0.5	0.795607	0.778591	0.778817	0.778801
0.6	0.714257	0.697427	0.697727	0.697676
0.7	0.627760	0.612361	0.612726	0.612626
0.8	0.540031	0.527036	0.527453	0.527292
0.9	0.454598	0.444636	0.445088	0.444858
1.0	0.374384	0.367713	0.368181	0.367879

TABLE 5.2. Comparison of the Euler, modified Euler, and improved Euler formulas with the exact solution for $f(x, y) = -2xy$. It is clear that modified and improved Euler formulas give a more accurate result than the simple Euler method.

5.1.5 *Alternative Derivation of the Improved Euler Method*

The improved Euler method could also be obtained by another procedure that will be useful later when we derive more accurate methods. The idea is to substitute the RHS of Equation (5.10) for $y(x_0+h)$, the second argument of f in the last line of (5.10). This yields

$$y(x_0 + h) = y_0 + \frac{h}{2}\{f_0 + f(x_0 + h, y(x_0) + \tfrac{h}{2}[f_0 + f(x_0 + h, y(x_0 + h))])\}$$

Next we approximate the embedded function $f(x_0 + h, y(x_0 + h))$ by a linear combination of the lower-order terms to which the function is being added. In the case at hand, $f(x_0 + h, y(x_0 + h))$ is added to f_0. So, "linear combination" means simply a constant multiple of f_0, and we write

$$f(x_0 + h, y(x_0 + h)) = \alpha f_0$$

This yields

$$y(x_0 + h) = y_0 + \frac{h}{2}\{f_0 + f(x_0 + h, y_0 + (h/2)[(1 + \alpha) f_0])\} \tag{5.12}$$

The constant α is determined by expanding both sides of this equation in a Taylor series up to the accuracy of the RHS (in this case, h^2) and equating the coefficients on both sides. For the LHS, we have

$$\begin{aligned}
\text{LHS} = y(x_0 + h) &= y_0 + hy'(x_0) + \frac{h^2}{2} y''(x_0) \\
&= y_0 + hf(x_0, y_0) + \frac{h^2}{2}[\partial_1 f(x_0, y_0) + f(x_0, y_0)\partial_2 f(x_0, y_0)] \\
&\equiv y_0 + hf_0 + \frac{h^2}{2}[\partial_1 f_0 + f_0 \partial_2 f_0] \qquad (5.13)
\end{aligned}$$

because

$$y''(x) = \frac{dy'}{dx} = \frac{df}{dx} = \frac{\partial f}{\partial x} + \frac{\partial f}{\partial y}\frac{dy}{dx} = \partial_1 f + y'\partial_2 f = \partial_1 f + f\partial_2 f$$

where ∂_k means partial derivative with respect to the kth variable. For the RHS—once we use the two-variable Taylor expansion for $f(x, y)$—we obtain

$$\begin{aligned} \text{RHS} &= y_0 + \frac{h}{2}f_0 + \frac{h}{2}\left[f_0 + h\partial_1 f_0 + \frac{h}{2}(1+\alpha)f_0\partial_2 f_0\right] \\ &= y_0 + hf_0 + \frac{h^2}{2}\left[\partial_1 f_0 + \frac{1+\alpha}{2}\partial_2 f_0\right] \end{aligned}$$

Comparison of this expression with (5.13) gives $\alpha = 1$. It follows from Equation (5.12) that

$$y(x_0 + h) = y_0 + \frac{h}{2}[f_0 + f(x_0 + h, y_0 + hf_0)]$$

which is the improved Euler method.

5.2 The Kutta Method

We can continue improving the integration and, consequently, the method of solving ODEs. For example, let us apply Simpson's rule to the integral in (5.3). Equation (4.6) then yields

$$\begin{aligned} y(x_0 + h) = y(x_0) + \int_{x_0}^{x_0+h} f(t, y(t))\,dt &\approx y(x_0) + \frac{h}{6}[f(x_0, y_0) \\ &+ 4f(x_0 + h/2, y(x_0 + h/2)) + f(x_0 + h, y(x_0 + h))] \\ &= y_0 + \frac{h}{6}[f_0 + 4z_1 + z_2] \end{aligned} \qquad (5.14)$$

with

$$\begin{aligned} z_1 &\equiv f(x_0 + h/2, y(x_0 + h/2)) \approx f(x_0 + h/2, y_0 + h/2f_0) \qquad (5.15) \\ z_2 &\equiv f(x_0 + h, y(x_0 + h)) = f(x_0 + h, y_0 + \tfrac{h}{6}[f_0 + 4z_1 + z_2]) \end{aligned}$$

where we substituted the RHS of (5.14) for the second argument of z_2.

Following the procedure outlined in the alternate derivation of the improved Euler method, we write $z_2 = \alpha f_0 + \beta z_1$ and substitute this on the RHS of the second equation of (5.15). Then Equation (5.14) becomes

$$\begin{aligned} y(x_0 + h) &= y_0 + \frac{h}{6}[f_0 + 4z_1 + f(x_0 + h, y_0 + \tfrac{h}{6}[(1+\alpha)f_0 + (4+\beta)z_1])] \\ &= y_0 + \frac{h}{6}[f_0 + 4z_1 + f(x_0 + h, y_0 + \tfrac{h}{6}Z)] \end{aligned} \qquad (5.16)$$

where $Z = (1+\alpha)f_0 + (4+\beta)z_1$.

We want to approximate the RHS of (5.16) in such a way that the Taylor expansions of both its sides agree up to h^2. The expansion of the LHS is given by Equation (5.13). To obtain the expansion for the RHS, we need the Taylor series of z_1 up to order h. This is

$$z_1 \approx f(x_0 + h/2, y_0 + h/2 f_0) = f_0 + \frac{h}{2}\partial_1 f_0 + \frac{h}{2} f_0 \partial_2 f_0$$

Similarly,

$$\begin{aligned} f(x_0 + h, y_0 + \tfrac{h}{6}Z) &\approx f_0 + h\partial_1 f_0 + \tfrac{h}{6} Z \partial_2 f_0 \\ &= f_0 + h\partial_1 f_0 + \tfrac{h}{6}[(1+\alpha)f_0 + (4+\beta)z_1]\partial_2 f_0 \end{aligned}$$

So, the RHS of (5.16) is

$$\begin{aligned} \text{RHS} &= y_0 + \frac{h}{6}\Big[f_0 + \overbrace{4f_0 + 2h\partial_1 f_0 + 2hf_0\partial_2 f_0}^{=4z_1} + f_0 + h\partial_1 f_0 \\ &\qquad + \tfrac{h}{6}[(1+\alpha)f_0 + (4+\beta)z_1]\partial_2 f_0\Big] \\ &= y_0 + hf_0 + \frac{h^2}{2}\partial_1 f_0 + \frac{h^2}{3} f_0\partial_2 f_0 + \frac{h^2}{36}[(1+\alpha) + (4+\beta)] f_0 \partial_2 f_0 \end{aligned}$$

where in the last step, we approximated z_1 by f_0 because the term was being multiplied by $h^2/36$ already. Comparison of the last equation with (5.13) shows that we must require

$$1 + \alpha + 4 + \beta = 6 \;\Rightarrow\; \alpha + \beta = 1$$

The two unknowns are *not* determined uniquely. Nevertheless, we have

$$z_2 = f(x_0 + h, y_0 + (h/6)[(1+\alpha)f_0 + (5-\alpha)z_1])$$

It is convenient to cancel out the factor of 6 in the denominator. The obvious choices $\alpha = 5$ (making the coefficient of f_0 equal to 6) or $\alpha = -1$ (making the coefficient of z_1 equal to 6) will eliminate either z_1 or f_0. A choice that retains both terms and gets rid of the factor 6 is $\alpha = -7$. Then

$$z_2 = f(x_0 + h, y_0 + h(2z_1 - f_0))$$

We can now generalize Equation (5.14) to

$$y_{k+1} \approx y_k + \frac{h}{6}\left(f_k + 4z_1^{(k)} + z_2^{(k)}\right) \tag{5.17}$$

where

$$z_1^{(k)} = f\left(x_k + \tfrac{1}{2}h, y_k + \tfrac{1}{2}hf_k\right), \qquad z_2^{(k)} = f\left(x_k + h, y_k + h(2z_1^{(k)} - f_k)\right)$$

Equation (5.17) is called the **Kutta method**. It is more accurate than any of the Euler methods.

Let us use the Kutta method to numerically solve the two DEs of Section 5.1.4. Aside from the initial data and the calculation of the independent variable array, which are unchanged, for the Kutta method, we have to type in Equation (5.17):

```
In[1]:= Kutta[h_,n_] := Do[yKut[i+1] =yKut[i]
        +(h/6)(f[x[i],yKut[i]]+4(f[x[i]+0.5 h,
        yKut[i]+0.5 h f[x[i],yKut[i]]])+f[x[i]+h,
        yKut[i]+h (2 (f[x[i]+0.5 h,yKut[i]+0.5 h f[x[i],
        yKut[i]]])-f[x[i],yKut[i]])]),{i,0,n}];
```

After running the input lines 1, 2, 6, and 7 of Section 5.1.4, as well as `In[1]` of this section, we type in

```
In[2]:= Kutta[0.05, 20];
```

to generate the new array of dependent variables for $f(x, y) = y$. Then we store the array so obtained in another array called y_1 as follows:

```
In[3]:= Do[y1[i] = yKut[i], {i,0,20,2}];
```

Now we change the function $f(x, y)$ to $-2xy$, run `in[1]` and `in[2]` above once more, and store the resulting array in y_2:

```
In[4]:= Do[y2[i] = yKut[i], {i,0,20,2}];
```

We can now display the results in a table for comparison with Tables 5.1 and 5.2. To do so, we type in

```
In[5]:= Table[{x[i],y1[i],Exp[x[i]],y2[i]
        Exp[-x[i]^2]},{i,0,20,2}]// MatrixForm
```

The output is a matrix with entries given in Table 5.3. Comparison of this table with Tables 5.1 and 5.2 clearly shows that the Kutta method is by far superior to any of the Euler methods.

5.3 The Runge–Kutta Method

Our alternative derivation of the improved Euler method and the derivation of the Kutta method point to a general procedure of obtaining more and more accurate techniques for solving DEs. We noted in those derivations that the arguments of the function $f(x, y)$ appearing in the DE were evaluated at different points $(x_0 + rh, y_0 + sh)$ where r and s were numbers between 0 and 1. A powerful method, due to Runge and Kutta and others, has been developed that gives extremely accurate results.

x	y1[i]	e^x	y2[i]	e^{-x^2}
0.0	1	1	1	1
0.1	1.10517	1.10517	0.990052	0.990050
0.2	1.22140	1.22140	0.960794	0.960789
0.3	1.34986	1.34986	0.913937	0.913931
0.4	1.49182	1.49182	0.852151	0.852144
0.5	1.64872	1.64872	0.778809	0.778801
0.6	1.82211	1.82212	0.697685	0.697676
0.7	2.01375	2.01375	0.612634	0.612626
0.8	2.22553	2.22554	0.527299	0.527292
0.9	2.45959	2.45960	0.444863	0.444858
1.0	2.71827	2.71828	0.367881	0.367879

TABLE 5.3. The result of applying the Kutta method to solve $y' = f(x, y)$. The second column gives the solution for $f(x, y) = y$, and the fourth column the solution for $f(x, y) = -2xy$.

The **Runge–Kutta method** starts with

$$y_{k+1} = y_k + h \left[\alpha_0 f(x_k, y_k) + \sum_{j=1}^{p} \alpha_j f(x_k + \mu_j h, y_k + b_j h) \right] \tag{5.18}$$

where α_0 and $\{\alpha_j, b_j, \mu_j\}_{j=1}^{p}$ are constants chosen so that if the RHS of (5.18) were Taylor-expanded in powers of h, the coefficients of a certain number of the leading terms would agree with the corresponding expansion coefficients of the Taylor expansion of the LHS. It is customary to express the b's as linear combinations of preceding values of f:

Runge–Kutta method

$$hb_i = \sum_{r=0}^{i-1} \lambda_{ir} z_r, \qquad i = 1, 2, \ldots, p$$

The z_r are recursively defined as

$$z_0 = hf(x_k, y_k), \qquad z_r = hf(x_k + \mu_r h, y_k + b_r h)$$

Then Equation (5.18) becomes

$$y_{k+1} = y_k + \sum_{r=0}^{p} \alpha_r z_r \tag{5.19}$$

The (nontrivial) task now is to determine the parameters α_r, μ_r, and λ_{ij}.

In general, the determination of these constants is extremely tedious. Let us consider the very simple case where $p = 1$, and let $\lambda \equiv \lambda_{01}$ and $\mu \equiv \mu_1$. Then $hb_1 = \lambda_{01} z_0 = \lambda z_0$, and we obtain

$$y_{k+1} = y_k + \alpha_0 z_0 + \alpha_1 z_1 \tag{5.20}$$

where $z_0 = hf(x_k, y_k)$ and

$$z_1 = hf(x_k + \mu h, y_k + hb_1) = hf(x_k + \mu h, y_k + \lambda z_0)$$

Taylor-expanding z_1, a function of two variables, gives[2]

$$\begin{aligned} z_1 &= hf(x_k, y_k) + h^2(\mu f_x(x_k, y_k) + \lambda f(x_k, y_k) f_y(x_k, y_k)) \\ &+ \frac{h^3}{2}\Big[\mu^2 f_{xx}(x_k, y_k) + 2\lambda\mu f(x_k, y_k) f_{xy}(x_k, y_k) \\ &+ \lambda^2 f^2(x_k, y_k) f_{yy}(x_k, y_k)\Big] + O(h^4) \end{aligned}$$

where $f_x \equiv \partial f/\partial x$ and $f_{xy} \equiv \partial^2 f/\partial x \partial y$, etc. Substituting this in (5.20) and using notations such as f_k, $f_{x,k}$, $f_{xx,k}$ for evaluation of f and its derivatives at (x_k, y_k), we get

$$\begin{aligned} y_{k+1} &= y_k + h(\alpha_0 + \alpha_1) f_k + h^2 \alpha_1 (\mu f_{x,k} + \lambda f_k f_{y,k}) \\ &+ \frac{h^3}{2}\alpha_1(\mu^2 f_{xx,k} + 2\lambda\mu f_k f_{xy,k} + \lambda^2 f_k^2 f_{yy,k}) + O(h^4) \end{aligned} \tag{5.21}$$

On the other hand, with

$$\begin{aligned} y' &= f, \qquad y'' = \frac{dy}{dx} = \frac{\partial y}{\partial x}\frac{dy}{dx} + \frac{\partial f}{\partial x} = y' f_y + f_x = f f_y + f_x \\ y''' &= f_{xx} + 2 f f_{xy} + f^2 f_{yy} + f_y(f f_y + f_x) \end{aligned}$$

the Taylor expansion of $y(x_k + h)$ gives

$$\begin{aligned} y_{k+1} &= y_k + h f_k + \frac{h^2}{2}(f_k f_{y,k} + f_{x,k}) \\ &+ \frac{h^3}{6}[f_{xx,k} + 2 f_k f_{xy,k} + f_k^2 f_{yy,k} + f_{y,k}(f_k f_{y,k} + f_{x,k})] + O(h^4) \end{aligned} \tag{5.22}$$

If we demand that (5.21) and (5.22) agree up to the h^2 term (we cannot demand agreement for h^3 or higher because of overspecification), then we must have $\alpha_0 + \alpha_1 = 1$, $\alpha_1 \mu = \frac{1}{2}$, $\alpha_1 \lambda = \frac{1}{2}$. There are only three equations for four unknowns. Therefore, there will be an arbitrary parameter β in terms of which the unknowns can be written:

$$\alpha_0 = 1 - \beta, \qquad \alpha_1 = \beta, \qquad \mu = \frac{1}{\beta}, \qquad \lambda = \frac{1}{2\beta}$$

Substituting these values in Equation (5.20) gives

$$y_{k+1} = y_k + h\left[(1-\beta) f_k + \beta f\left(x_k + \frac{h}{2\beta}, y_k + \frac{h f_k}{2\beta}\right)\right] + O(h^3)$$

---

[2]Recall from Section 4.3 that the symbol $O(h^m)$ means that all terms of order h^m and higher have been neglected.

This formula becomes useful if we let $\beta = \frac{1}{2}$. Then $x_k + h/2\beta = x_k + h$, which makes the evaluation of the second term in square brackets convenient. For $\beta = \frac{1}{2}$, we have

$$y_{k+1} = y_k + \frac{h}{2}[f_k + f(x + h, y_k + hf_k)] + O(h^3) \tag{5.23}$$

which the reader recognizes as the improved Euler formula.

Formulas that give more accurate results can be obtained by retaining terms beyond $p = 1$. Such calculations are extremely tedious, and we shall not reproduce them here, being content with the final results. For $p = 2$, if we write

$$y_{k+1} = y_k + \sum_{r=0}^{2} \alpha_r z_r = y_k + \alpha_0 z_0 + \alpha_1 z_1 + \alpha_2 z_2$$

there will be eight unknowns (three α's, three λ_{ij}'s, and two μ's), and the demand for agreement between the Taylor expansion and the expansion of f up to h^3 will yield only six equations. Therefore, there will be two arbitrary parameters whose specification results in various formulas. One such formula is the Kutta formula (5.17), which we have encountered before. A second formula, due to Heun, has the form

Heun formula

$$y_{k+1} = y_k + \tfrac{1}{4}(z_0 + 3z_2) + O(h^4)$$

where

$$z_0 = hf(x_k, y_k), \qquad z_1 = hf(x_k + \tfrac{1}{3}h, y_k + \tfrac{1}{3}z_0)$$
$$z_2 = hf(x_k + \tfrac{2}{3}h, y_k + \tfrac{2}{3}z_1 - z_0)$$

Kutta and Heun formulas are of about the same order of accuracy. What is nice about all such formulas is that we can plug in the known quantities x_k and y_k—starting with $k = 0$—on the RHS and find y_{k+1}.[3]

The accuracy of the Runge–Kutta method and the fact that it requires no startup procedure (i.e., all quantities on the RHS of the y_{k+1} equation are known) make it at once readily usable in computer programs and one of the most popular methods for solving differential equations.

The Runge–Kutta method can be made more accurate by using higher values of p. For instance, a formula used for $p = 3$ is

$$y_{k+1} = y_k + \tfrac{1}{6}(z_0 + 2z_1 + 2z_2 + z_3) + O(h^5) \tag{5.24}$$

---

[3]Other—less accurate—methods, which we have not discussed, have quantities on the RHS whose values are to be found using other techniques of approximation. See [Hass 99], Chapter 13, for a discussion of these methods.

where

$$z_0 = hf(x_k, y_k), \qquad z_1 = hf(x_k + \tfrac{1}{2}h, y_k + \tfrac{1}{2}z_0)$$
$$z_2 = hf(x_k + \tfrac{1}{2}h, y_k + \tfrac{1}{2}z_1), \qquad z_3 = hf(x_k + h, y_k + z_2)$$

Equation (5.24) is *the* Runge–Kutta formula and is the main ammunition in the arsenal of numerical solution of DEs.

5.4 Higher-Order Equations

Any nth-order differential equation is equivalent to n first-order differential equations in $n+1$ variables. Thus, for instance, the most general second-order DE, $y'' = g(x, y, y')$, can be reduced to two first-order DEs by defining $y' = u$ and writing the original DE as the system of equations

$$u' = g(x, y, u), \qquad y' = u$$

These two equations are completely equivalent to the original second-order DE. Thus, it is appropriate to discuss numerical solutions of systems of first-order DEs in several variables. The discussion here is limited to systems consisting of two equations. The generalization to several equations is not difficult.

Consider the following system of coupled equations with the indicated initial conditions:

$$y' = f(x, y, u), \ y(x_0) = y_0, \qquad u' = g(x, y, u), \ u(x_0) = u_0 \tag{5.25}$$

Using an obvious generalization of Equation (5.24), we can write

$$\begin{aligned} y_{k+1} &= y_k + \tfrac{1}{6}(z_0 + 2z_1 + 2z_2 + z_3) + O(h^5) \\ u_{k+1} &= u_k + \tfrac{1}{6}(w_0 + 2w_1 + 2w_2 + w_3) + O(h^5) \end{aligned} \tag{5.26}$$

where

$$z_0 = hf(x_k, y_k, u_k), \qquad z_1 = hf(x_k + \tfrac{1}{2}h, y_k + \tfrac{1}{2}z_0, u_k + \tfrac{1}{2}w_0)$$
$$z_2 = hf(x_k + \tfrac{1}{2}h, y_k + \tfrac{1}{2}z_1, u_k + \tfrac{1}{2}w_1)$$
$$z_3 = hf(x_k + h, y_k + z_2, u_k + w_2)$$

and

$$w_0 = hg(x_k, y_k, u_k), \qquad w_1 = hg(x_k + \tfrac{1}{2}h, y_k + \tfrac{1}{2}z_0, u_k + \tfrac{1}{2}w_0)$$
$$w_2 = hg(x_k + \tfrac{1}{2}h, y_k + \tfrac{1}{2}z_1, u_k + \tfrac{1}{2}w_1)$$
$$w_3 = hg(x_k + h, y_k + z_2, u_k + w_2)$$

These formulas are more general than needed for a second-order DE, since, as mentioned above, such a DE is equivalent to the simpler system in which $f(x, y, u) \equiv u$. Therefore, with

$$z_0 = hu_k, \qquad z_1 = h(u_k + \tfrac{1}{2}w_0)$$
$$z_2 = hu_k + \tfrac{1}{2}hw_1, \qquad z_3 = hu_k + hw_2$$

Equation (5.26) specializes to

Numerical solution of a second-order DE

$$\begin{aligned} y_{k+1} &= y_k + hu_k + \tfrac{1}{6}h(w_0 + w_1 + w_2) + O(h^5) \\ u_{k+1} &= u_k + \tfrac{1}{6}(w_0 + 2w_1 + 2w_2 + w_3) + O(h^5) \end{aligned} \tag{5.27}$$

where

$$\begin{aligned} w_0 &= hg(x_k, y_k, u_k), \qquad w_1 = hg(x_k + \tfrac{1}{2}h, y_k + \tfrac{1}{2}hu_k, u_k + \tfrac{1}{2}w_0) \\ w_2 &= hg(x_k + \tfrac{1}{2}h, y_k + \tfrac{1}{2}hu_k + \tfrac{1}{4}hw_0, u_k + \tfrac{1}{2}w_1) \\ w_3 &= hg(x_k + h, y_k + hu_k + \tfrac{1}{2}hw_1, u_k + w_2) \end{aligned}$$

Equation (5.27) is especially suitable for initial-value problems in which the solution and its derivative are specified at some initial time (usually $t = 0$), and then subsequently *both* are calculated at later times (recall that u is the derivative of y). Because of such determination of both the solution and its derivative, Equation (5.27) is also useful in constructing **phase-space diagrams**, a plot of the solution (usually the horizontal axis) versus its derivative (usually the vertical axis).

phase-space diagram defined

Since there are two coupled equations in y and u, we cannot use the `Do` command. The suitable procedure is using `For` discussed in Section 3.2. So, we type in[4]

```
In[1]:= RungKut[x0_, y0_, u0_, h_, n_] :=
        For[i=0; x[0]=x0; y[0]=y0; u[0]=u0, i<=n, i=i+1,
        x[i]=x0+i h; w0=h g[x[i],y[i],u[i]];
        w1=h g[x[i]+0.5 h,y[i]+0.5 h u[i],u[i]+0.5 w0];
        w2=h g[x[i]+0.5 h,y[i] 0.5 h u[i]+0.25 h w0,
            u[i]+0.5 w1];
        w3=h g[x[i]+h,y[i]+h u[i]+0.5 h w1,u[i]+w2];
        y[i+1]=y[i]+h u[i]+(h/6)(w0+w1+w2);
        u[i+1]=u[i]+(1/6)(w0+2 w1+2 w2+w3)]
```

Once we know the function $g(x, y, u)$ and the initial data, we can calculate the array `y[i]`.

As a specific example, let us solve the very simple DE $y'' + y = 0$ with the initial conditions $y(0) = 0$, $y'(0) = 1$. It should be obvious that the

[4]We could have used `For` instead of `Do` in Sections 5.1.4 and 5.2 as well.

analytic solution of this DE is $y(x) = \sin x$. The idea is to test the accuracy of the Runge–Kutta method. For this problem $g(x, y, u) = -y$. So, we type in

```
In[2]:= g[x_,y_,u_]:=-y
```

If we use the same step sizes as in the previous sections, we will not see any difference between the Runge–Kutta solution and the exact solution. So, first let us take the step size to be 0.1 and store every other value of the solution in an array—of length 11—named `yAccurate`. We do this by typing in

```
In[3]:= RungKut[0,0,1,0.1,20];
```

and

```
In[4]:= Do[yAccurate[i]=y[2i], {i,0,10}];
```

Now increase the step size to 0.2, and recalculate the—less accurate—array (also of length 11):

```
In[5]:= RungKut[0,0,1,0.2,10];
```

To display the two solutions, and compare them with the exact result, we type in

```
In[6]:= Table[{x[i],y[i],yAccurate[i],Sin[x[i]]},
        {i,1,10}]//MatrixForm
```

The output will be a matrix with entries as given in Table 5.4. Even for the fairly large step size of 0.1, there is hardly any noticeable difference between the Runge–Kutta solution and the exact solution. Only when we increase the step size to 0.2, do we notice differences; differences that are noticeably smaller than any of the Euler methods, *even when* $h = 0.05$ *for the latter*!

5.5 Eigenvalue Problems

MM, Section 11.1

When partial differential equations of mathematical physics are separated into ODEs, they most often lead to a second-order linear DE known as a *Sturm–Liouville system.* The most general Sturm–Liouville system is of the form

$$f_2(x)y''(x) + f_1(x)y'(x) + f(x)y(x) = \lambda y(x) \tag{5.28}$$

where *both* $y(x)$, called the *eigenfunction* of the system, and λ, called its *eigenvalue*, are to be determined subject to appropriate boundary conditions (BCs). In this context, the Sturm–Liouville system is also referred to as an *eigenvalue problem.*

x[i]	y[i]	yAccurate[i]	Sin[x[i]]
0	0	0	0
0.2	0.198667	0.198669	0.198669
0.4	0.389413	0.389418	0.389418
0.6	0.564635	0.564642	0.564642
0.8	0.717347	0.717356	0.717356
1.0	0.841462	0.841470	0.841471
1.2	0.932031	0.932039	0.932039
1.4	0.985444	0.985449	0.985445
1.6	0.999571	0.999574	0.999574
1.8	0.973849	0.973848	0.973848
2.0	0.909304	0.909298	0.909297

TABLE 5.4. Comparison of the Runge–Kutta and exact solutions to the second-order DE $y'' = -y$ with $x_0 = y_0 = 0$ and $y'_0 = 1$. The second and third columns give values for the solution when $h = 0.2$ and $h = 0.1$, respectively.

There are various numerical techniques for finding the eigenvalues and eigenvectors of an eigenvalue problem. We shall use the direct and general method of *discretization*, whereby the DE is turned into a matrix equation. To do this, we first have to approximate the derivative with a discrete formula.

5.5.1 *Discrete Differentiation*

Any discrete quantity is an approximation to its continuous counterpart. The art of numerical analysis is to make this approximation as close to the exact value as possible while keeping the discrete expressions as simple as possible. Let us begin by writing two Taylor expansions of an arbitrary function f:

$$\begin{aligned} f(x+h) &= f(x) + hf'(x) + \frac{h^2}{2!}f''(x) + \frac{h^3}{3!}f'''(x) + \frac{h^4}{4!}f^{\text{iv}}(x) + \dots \\ f(x-h) &= f(x) - hf'(x) + \frac{h^2}{2!}f''(x) - \frac{h^3}{3!}f'''(x) + \frac{h^4}{4!}f^{\text{iv}}(x) + \dots \end{aligned} \tag{5.29}$$

We rewrite the first equation as

$$f'(x) = \frac{f(x+h) - f(x)}{h} \underbrace{- \frac{h}{2!}f''(x) - \frac{h^2}{3!}f'''(x) - \frac{h^3}{4!}f^{\text{iv}}(x) + \dots}_{\text{this remainder is denoted by } O(h)}$$

The symbol $O(h)$ (read *order of* h) tells us that the largest term in the remainder, which is the term with the lowest power of h, is multiplied by h to the first power. It is the same "O" function encountered in Sections

4.3 and 5.3. The long expression above can be shortened to

$$f'(x) = \frac{f(x+h) - f(x)}{h} + O(h)$$

indicating that the neglected terms were "of order h." A good numerical expression is that in which the neglected terms are of the highest possible order in h: the higher the power of h in the largest term of the neglected expression, the smaller the error. This is, of course, because h is a small number in all numerical calculations. Can we come up with a formula for the derivative with $O(h^2)$? If we subtract the second series in Equation (5.29) from the first, we get

$$f(x+h) - f(x-h) = 2hf'(x) + \frac{h^3}{3}f'''(x) + \dots$$

or

$$f'(x) = \frac{f(x+h) - f(x-h)}{2h} - \frac{h^2}{6}f'''(x) - \dots$$

or

$$f'(x) = \frac{f(x+h) - f(x-h)}{2h} + O(h^2) \tag{5.30}$$

Thus, this *central difference* formula for the derivative, which we shall use hereafter, is more accurate than the previous formula. One can obtain more accurate expressions for the derivative with $O(h^4)$ for the remainder, but the formulas become much more complicated. We shall stick to Equation (5.30) for the first derivative.

The eigenvalue problem is a DE of the second order. So, we need the second derivative as well. To obtain an expression for the second derivative, we add the two series of (5.29). This yields

$$f(x+h) + f(x-h) = 2f(x) + h^2 f''(x) + \frac{h^4}{12}f^{\mathrm{iv}}(x) + \dots$$

or

$$f''(x) = \frac{f(x+h) - 2f(x) + f(x-h)}{h^2} - \frac{h^2}{12}f^{\mathrm{iv}}(x) - \dots$$

or

$$f''(x) = \frac{f(x+h) - 2f(x) + f(x-h)}{h^2} + O(h^2) \tag{5.31}$$

In terms of indices, the first and second derivatives are written as

$$\begin{aligned} \mathbf{D}f_i &= \frac{f_{i+1} - f_{i-1}}{2h} + O(h^2) \\ \mathbf{D}^2 f_i &= \frac{f_{i+1} - 2f_i + f_{i-1}}{h^2} + O(h^2) \end{aligned} \tag{5.32}$$

5.5.2 Discrete Eigenvalue Problem

We can now tackle the problem of solving our Sturm–Liouville system numerically. Before doing so, we first simplify the DE. Specifically, we will eliminate the first derivative in Equation (5.28). This can be done by defining a new function $u(x)$ via

$$y(x) = w(x)u(x) \quad \text{where} \quad w(x) = \exp\left(-\tfrac{1}{2}\int_a^x \frac{f_1(t)}{f_2(t)}\,dt\right) \tag{5.33}$$

and a is a convenient constant. Then, the reader may show that $u(x)$ satisfies the following DE:

$$f_2 u'' + \left(f - \frac{f_1^2 + 2f_1' f_2 - 2f_2' f_1}{4f_2}\right) u = \lambda u \tag{5.34}$$

So, using different symbols for the coefficient functions, we can assume that our DE is of the form

$$p(x)y''(x) + q(x)y(x) = \lambda y(x) \quad \text{or} \quad p_i \mathbf{D}^2 y_i + q_i y_i = \lambda y_i \tag{5.35}$$

where the second equation is the discretized version of the DE and the one we want to solve.

Substituting the discretized second derivative from Equation (5.32) and rearranging, we obtain the following *master* equation:

$$p_i y_{i+1} + (h^2 q_i - 2p_i) y_i + p_i y_{i-1} = h^2 \lambda y_i \tag{5.36}$$

The solution to this equation, i.e., the determination of λ and the set of y_i, requires some boundary conditions. We shall look at some specific examples in the next chapter.

5.6 Problems

Problem 5.1. Choose some values for γ and k and some initial conditions (initial values for x and $\dot{x}$), and use
(a) one of Euler's methods and
(b) the Runge–Kutta method to find a numerical solution to the damped harmonic oscillator DE

$$\ddot{x} + \gamma\dot{x} + kx = 0$$

where the dot indicates differentiation with respect to time. Plot the solution as a function of time and compare it with the plot of the known analytic solution, making sure the time interval of motion is long enough to observe a few oscillations of the system.

Problem 5.2. The undamped, undriven pendulum obeys the DE

$$\ddot{\theta} + \sin\theta = 0$$

where θ is the angle the pendulum makes with the vertical, and the dot indicates differentiation with respect to time. Assume that $\theta(0) = \pi/2$ and $\dot{\theta}(0) = 0$.
(a) Use the simplest Euler method and Runge–Kutta method to find a numerical solution for this DE. Compare the results.
For the rest of the problem concentrate on the Runge–Kutta solution only.
(b) Plot the Runge–Kutta solution as a function of time. Make sure the time interval of motion is long enough to observe a few oscillations of the system.
(c) Using `ParametricPlot`, plot the phase-space diagram (see page 169) with $\dot{\theta}(t)$ as the vertical axis and $\theta(t)$ as the horizontal axis for $\theta(0) = \pi/10$ and $\dot{\theta}(0) = 0$.
(d) Plot a second phase-space diagram for $\theta(0) = 179.98°$ and $\dot{\theta}(0) = 1$.

Problem 5.3. The damped, driven pendulum obeys the DE

$$\ddot{\theta} + \gamma\dot{\theta} + \sin\theta = \psi_0 \cos(\omega t)$$

where θ is the angle the pendulum makes with the vertical, ω is the driving frequency, ϕ_0 is some constant, and the dot indicates differentiation with respect to time.
(a) Use the Runge–Kutta method to find a numerical solution for this DE.
(b) Plot the phase-space diagram (see page 169) for $\gamma = 0.3$, $\phi_0 = 0.5$, $\omega = 1$, $\theta(0) = \pi/1.0001$, $\dot{\theta}(0) = 0$, and $0 \le t \le 100$.
(c) Change ϕ_0 and ω in (b) to 1.15 and 2/3, respectively—keeping all the other parameters the same—and replot the phase-space diagram.

Problem 5.4. The *Lorenz* DE is

$$\begin{aligned}
\dot{x}(t) &= \sigma[-x(t) + y(t)] \\
\dot{y}(t) &= rx(t) - y(t) - x(t)z(t) \\
\dot{z}(t) &= x(t)y(t) - qz(t)
\end{aligned}$$

where σ, r, and q are constants.
(a) Choose $\sigma = 10$, $r = 76$, and $q = 9$, and for some initial conditions of your choice use the Runge–Kutta method to find a numerical solution for this Lorenz DE.
(b) Plot $y(t)$ versus $x(t)$ using `ParametricPlot`.
(c) For the same σ, r, q, and initial conditions as in (a), plot $z(t)$ versus $x(t)$ using `ParametricPlot`.
(d) For the same σ, r, q, and initial conditions as in (a), make a three-dimensional parametric plot of the solution.

Problem 5.5. The DE governing the motion of a particle with electric charge q in an electromagnetic field is $m\ddot{\mathbf{r}} = q(\mathbf{E} + \mathbf{v} \times \mathbf{B})$, where m, $\mathbf{r}$, $\mathbf{v}$, $\mathbf{E}$, and $\mathbf{B}$ are, respectively, the mass of the particle, its position, its velocity, the electric field, and the magnetic field. This *Lorentz force law* can be written in component form as

$$\begin{aligned}
\ddot{x}(t) &= q(E_x + \dot{y}B_z - \dot{z}B_y) \\
\ddot{y}(t) &= q(E_y + \dot{z}B_x - \dot{x}B_z) \\
\ddot{z}(t) &= q(E_z + \dot{x}B_y - \dot{y}B_x)
\end{aligned}$$

Consider the case where the electric field is zero and the magnetic field is constant (set it equal to 1) along the z-axis. Suppose that the particle (of unit charge and unit mass) starts at the origin with $\dot{x}(0) = 1 = \dot{z}(0)$ and $\dot{y}(0) = 0$.
(a) Use the Runge–Kutta method to find a numerical solution for this set of DEs.
(b) Using `ParametricPlot3D`, find the trajectory of the particle for $0 \leq t \leq 30$.

Problem 5.6. Show that if one substitutes (5.33) in Equation (5.28) one gets (5.34). Use *Mathematica* to do all the differentiations.

6

Numerical Solutions of ODEs: Examples Using *Mathematica*

The techniques of Chapter 5, in particular, the Runge–Kutta method, are useful in solving differential equations that do not yield to analytic solutions, and aside from a handful of exceptions, those are precisely the type of DEs one encounters in practice. These methods have been used by programmers for decades and are still used in numerically intensive problems.

Mathematica, as a high-level computational software, is especially suited for solving DEs both analytically (when such solutions exist) and numerically. Furthermore, one can take advantage of the highly versatile `Graphics` package included in *Mathematica* to translate the solutions of DEs into sophisticated plots. In this chapter we examine a few DEs in great detail in the hope that the reader will glean from them the essentials of how to use *Mathematica* to solve DEs numerically.

6.1 Some Analytic Solutions

You can solve most of the elementary DEs with *Mathematica* using `DSolve`. For example,

using **DSolve**

```
In[1]:= DSolve[y''[x] + 4 y[x] == 0, y[x], x]
```

yields

```
Out[1]:=
```

$$\{\{y[x]->C[1]\,\mathrm{Cos}[2x]+C[2]\,\mathrm{Sin}[2x]\}\}$$

including two unknown constants of integration. Note that because DE is an "equation," *Mathematica* requires the double equality sign. We can also include some initial conditions as arguments of `DSolve`:

```
In[2]:= DSolve[{y''[x] + 4 y[x] == 0, y[0] == 0,
          y[Pi/4] == 1}, y[x], x]
```

and get

```
Out[2]:=
```

$$\{\{y[x]->\mathrm{Sin}[2x]\}\}$$

or specify only one initial condition

```
In[3]:= DSolve[{y''[x]+4 y[x]==0, y[0]==0}, y[x], x]
```

to get

```
Out[3]:=
```

$$\{\{y[x]->C[2]\,\mathrm{Sin}[2x]\}\}$$

Mathematica recognizes most of the special DEs of mathematical physics. Typing in the *standard* form of the Legendre DE,

MM, p. 533 and Box 12.3.2

```
In[4]:= DSolve[(1-x^2) y''[x]-2 x y'[x]+n(n+1) y[x]==0,
          y[x], x]
```

yields

```
Out[4]:=
```

$$\{\{y[x]->C[1]\,\mathrm{LegendreP}[n,x]+C[2]\,\mathrm{LegendreQ}[n,x]\}\}$$

MM, p. 564

and

```
In[5]:= DSolve[x^2 y''[x]+x y'[x]+(x^2-m^2) y[x]==0,
          y[x], x]
```

gives

```
Out[5]:=
```

$$\{\{y[x]->\mathrm{BesselJ}[m,x]C[1]+\mathrm{BesselY}[m,x]C[2]\}\}$$

However, *Mathematica* can also find the relation—if it exists—between a DE and the standard forms above. For example,

```
In[6]:= DSolve[(1+x^2) y''[x]-2 x y'[x]+a y[x]==0,
          y[x], x]
```

puts out

Out[6]:=

$$\{\{y[x]->(1+x^2)C[1]\,\text{LegendreP}\left[\tfrac{1}{2}\left(-1+\sqrt{9-4a}\right),2,ix\right]$$
$$+(1+x^2)C[2]\,\text{LegendreQ}\left[\tfrac{1}{2}\left(-1+\sqrt{9-4a}\right),2,ix\right]\}\}$$

where `LegendreP[n,m,x]` and `LegendreQ[n,m,x]` are the *associated Legendre functions*, denoted by $P_n^m(x)$ and $Q_n^m(x)$ in the mathematics literature. Similarly,

```
In[7]:= DSolve[x^2 y''[x]-x y'[x]-(x^2-m^2) y[x]==0,
          y[x], x]
```

yields

Out[7]:=

$$\{\{y[x]->x\,\text{BesselJ}\left[\sqrt{1-m^2},-ix\right]C[1]+x\,\text{BesselY}\left[\sqrt{1-m^2},-ix\right]C[2]$$

Almost all special functions of mathematical physics are related to the *hypergeometric function*, and *Mathematica* recognizes the standard DE satisfied by it,

MM, p. 275

```
In[8]:= DSolve[x(1-x) y''[x]-(c-(a+b+1)x) y'[x]
          -a b y[x]==0, y[x], x]
```

Out[8]:=

$$\{\{y[x]->C[1]\,\text{Hypergeometric2F1}[a,b,c,x]$$
$$+(-1)^{(1-c)}x^{(1-c)}C[2]\,\text{Hypergeometric2F1}[1+a-c,1+b-c,2-c,x]\}\}$$

and the nonstandard DE related to it:

```
In[9]:= DSolve[(ax+bx^2) y''[x]-(c-dx) y'[x]-e y[x]==0,
          y[x], x]
```

Out[9]:=

$$\{\{y[x]->C[1]\,\text{Hypergeometric2F1}[-\frac{1}{2}+\frac{d}{2b}-\frac{(b-d)^2-4be}{2b},$$
$$-\frac{1}{2}+\frac{d}{2b}+\frac{(b-d)^2-4be}{2b},\frac{c}{a},\frac{bx}{a}]$$
$$+(-1)^{\frac{a-c}{a}}a^{-\frac{a-c}{a}}b^{\frac{a-c}{a}}x^{\frac{a-c}{a}}$$
$$C[2]\,\text{Hypergeometric2F1}[\frac{1}{2}-\frac{c}{a}+\frac{d}{2b}-\frac{(b-d)^2-4be}{2b},$$
$$\frac{1}{2}-\frac{c}{a}+\frac{d}{2b}+\frac{(b-d)^2-4be}{2b},2-\frac{c}{a},\frac{bx}{a}]\}\}$$

As impressive as the list of differential equations familiar to *Mathematica* is, the real power of *Mathematica* shows up in the *numerical* solutions of DEs, which we illustrate next.

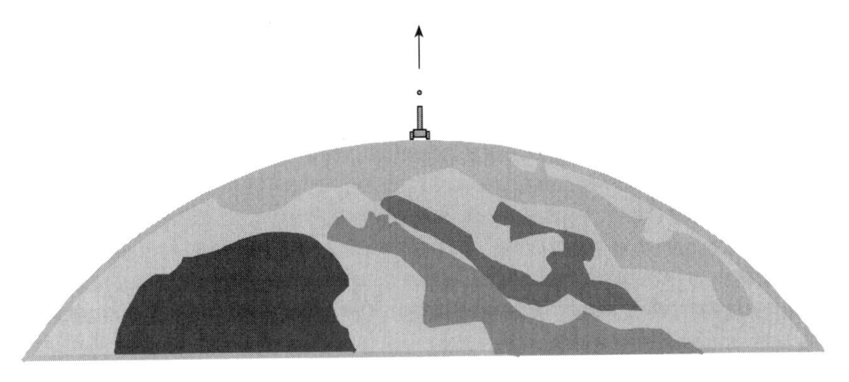

FIGURE 6.1. A projectile being fired vertically with arbitrary velocity.

6.2 A One-Dimensional Projectile

In this first example, we want to investigate the motion of a projectile fired straight up with some given velocity (Figure 6.1). We shall not put any restriction on the initial speed of the projectile, making it possible for it to reach heights that may be comparable to the Earth's radius. With such enormous distances possible, we can no longer assume that the weight of the projectile—and therefore, its acceleration—is constant, but we have to include variation of acceleration with distance.

It follows that the equation of motion of the projectile is

$$m\frac{d^2x}{dt^2} = -\frac{GmM}{(R+x)^2} \tag{6.1}$$

where m is the mass of the projectile, G is the universal gravitational constant, having the value 6.6726×10^{-11} in the International System of Units (SI), M is the mass of the Earth equal to 5.98×10^{24} kg, R is the Earth's radius equal to 6.37×10^6 m, and x is the height above the Earth's surface. The negative sign is due to the fact that we have taken "up" to be positive.

To make the problem more realistic, we want to include the air drag. For x small compared to the thickness of the atmosphere, the drag force is taken to be proportional to the velocity of the projectile. This approximation is good as long as the density of the atmosphere is uniform; and for small heights, this is indeed the case. However, we know that the atmospheric density decreases (exponentially) with height. To incorporate this variation, let us assume that the "constant" of proportionality multiplying the speed—to give the drag force—is an exponentially decreasing function of height. This is a reasonable assumption, because it is based on another reasonable assumption, namely that the drag coefficient is proportional to the density of the atmosphere. Including this force and dividing Equation

(6.1) by m, we obtain

$$\frac{d^2x}{dt^2} = -\frac{GM}{(R+x)^2} - ae^{-bx}\frac{dx}{dt} \tag{6.2}$$

where a is the drag coefficient at the surface of the Earth ($x = 0$), and b determines the rate at which this coefficient decreases with height.

Our task now is to find a solution to the DE (6.2) using *Mathematica*. So that we can use the equation for different situations, we take all terms of Equation (6.2) to the left-hand side, and define the expression `eqProj`, depending on the relevant parameters:

```
In[1]:= eqProj[G_, M_, R_, a_, b_] :=
        D[x[t],{t,2}]+G M/(R+x[t])^2+a E^(-b x[t])
        D[x[t], t]
```

Next, we solve the DE and give the process of solution a name:

```
In[2]:= proj[G_,M_,R_,a_,b_,x0_,v0_,T_] :=
        NDSolve[{eqProj[G,M,R,a,b]==0,
        x'[0]==v0, x[0]==x0}, x, {t,0,T}]
```

A couple of remarks are in order before we substitute actual numbers. First, `NDSolve` is *Mathematica*'s command for solving DEs numerically (thus, `N`). Second, the arguments of `NDSolve` consist of a list, followed by the dependent variable of the DE, then another list. The first list is a set of *equations*, starting with the DE, followed by *initial conditions*—as above—or boundary conditions in which the value of $x(t)$ is given at an appropriately chosen number of times. As a rule, an nth-order DE requires n initial (or boundary) conditions. Third, note that the initial (or boundary) conditions ought to be written in the form of *equations*.

With a (parametrized) solution at our disposal, we can look at some specific examples. First, let us consider the simplest case, where the drag force is zero and the initial velocity is small enough that the maximum height is negligible compared with the Earth's radius. This is the familiar case of constant acceleration with the simple solution

$$x(t) = v_0t + \tfrac{1}{2}gt^2$$

where $g = GM/R^2$ is the gravitational acceleration at the surface of the Earth, whose value for the parameters given above is $g = 9.8337103$. For $v_0 = 100$ m/s, we get

$$x(t) = 100t - 4.9168552t^2 \tag{6.3}$$

whose plot is shown in Figure 6.2.

To find the corresponding plot of the numerical solution, we type in

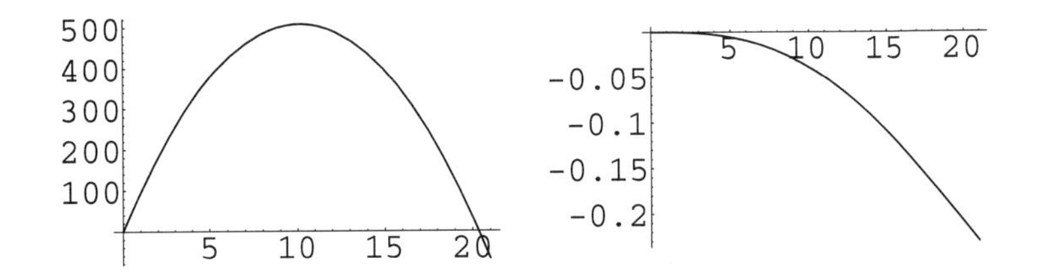

FIGURE 6.2. (Left) Plot of $x(t)$ versus t using Equation (6.3). (Right) Plot of the difference between Equation (6.3) and the solution obtained from numerical calculation.

```
In[3]:= projMotion=proj[6.6726 10^(-11), 5.98 10^(24),
        6.37 10^6, 0, 0, 0, 100, 25]
```

and to plot the result, we type in

```
In[4]:= Plot[x[t] /. projMotion, {t, 0, 21}]
```

and obtain a graph almost identical to the one given in Figure 6.2 on the left. To compare the two graphs more accurately, we plot the difference between the two functions by typing in

```
In[5]:= Plot[100 t-4.9168552 t^2-(x[t]/.projMotion),
        {t,0,21}]
```

The result is the plot given in Figure 6.2 on the right. The difference is remarkably small: for heights reaching hundreds of meters, the maximum difference is only 20 cm! The point we are trying to make is that the numerical solution is very accurate.

The next step in our treatment of the projectile is to include the drag force but still keep the heights small compared to the Earth's radius. It then follows that the variation of the drag coefficient is negligible, i.e., we can set $b = 0$. Then the equation of motion becomes

$$\frac{d^2x}{dt^2} = -g - a\frac{dx}{dt} \quad \text{or} \quad \frac{dv}{dt} = -g - av \quad \text{or} \quad \frac{dv}{g + av} = -dt$$

This can be easily integrated to give

$$v = \left(v_0 + \frac{g}{a}\right) e^{-at} - \frac{g}{a}$$

where the constant of integration has been written in terms of the initial speed v_0. Integration of the last equation gives $x(t)$:

$$x(t) = x_0 + \left(\frac{v_0}{a} + \frac{g}{a^2}\right)\left(1 - e^{-at}\right) - \frac{g}{a}t \tag{6.4}$$

Once again, we can compare the analytic and the numerical results. First we write the analytic result in *Mathematica*

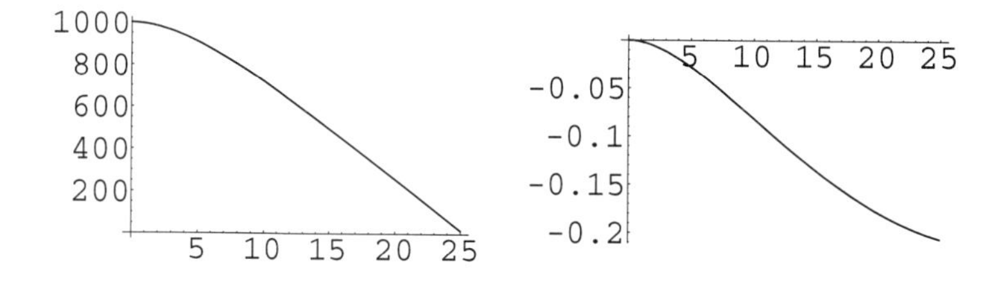

FIGURE 6.3. (Left) Plot of $x(t)$ versus t using Equation (6.4). (Right) Plot of the difference between Equation (6.4) and the solution obtained from numerical calculation.

```
In[6]:= xAnal[t_,x0_,v0_,a_,g_] :=
        x0+(v0/a+g/a^2) (1-E^(-a t))-(g/a) t
```

Then we take the case of a parachute landing with a drag coefficient of $a = 0.2$, in which $v_0 = 0$ and $x_0 = 1000$ m. So, the numerical solution is obtained by typing in

```
In[7]:= projMotion = proj[6.6726 10^(-11), 5.98 10^24,
        6.37 10^6, 0.2, 0, 1000, 0, 25]
```

Plotting either the analytic or the numerical solution, we get the graph on the left of Figure 6.3. Notice how after about 7 seconds, the graph of height becomes a straight line. This is the part of motion when the parachute reaches its *terminal velocity*, beyond which the motion has no acceleration. The graphs are too similar to be a good basis for comparison. However, if we plot the *difference* between the two functions by typing in

```
In[8]:= Plot[xAnal[t,1000,0,0.2,9.8337103]
        -(x[t]/.projMotion),{t,0,25}]
```

we obtain the plot on the right of Figure 6.3. The difference is indeed negligible!

Having gained confidence that the numerical solution is indeed valid, we can now attack the problem of large velocities and altitudes. As before, we first do the problem without drag. By typing in

```
In[1]:= projMotion=proj[6.6726 10^(-11), 5.98 10^(24),
        6.37 10^6, 0, 0, 0, 1000, 210];
        Plot[x[t] /. projMotion, {t, 0, 210}]
```

we obtain the plot in Figure 6.4. The entire motion takes about 200 s and the maximum height reached is approximately 50 km. Let us increase the initial speed to 5000 m/s, and from the plot (not shown) read off the duration of motion and the maximum height. These are 1350 s and 1.7×10^6 m. Continuing this process, we obtain 18,900 s for the duration

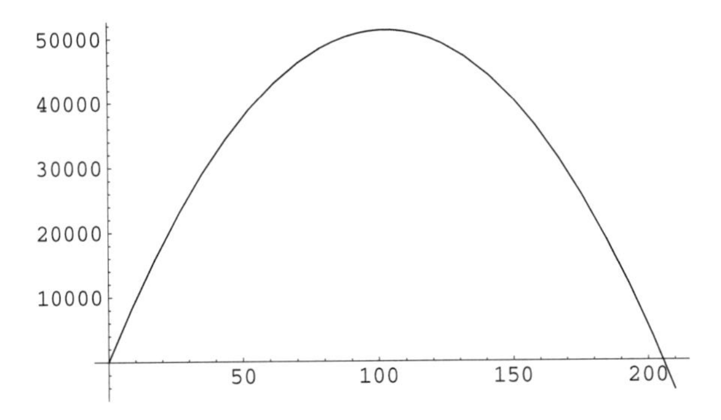

FIGURE 6.4. The graph of the dragless projectile motion when the initial speed is 1000 m/s.

and 2.52×10^7 m for the maximum height when the initial speed is 10,000 m/s. For the initial speed of 11,000 m/s, we get 282,000 s and 1.8×10^8 m or 78.3 hr and 28 Earth radii. And if you increase the initial speed to 11,500 m/s, no maximum will be reached because the graph will be very nearly a straight line. The projectile will be moving away forever with (almost) constant speed. This is because 11,500 m/s happens to be larger than the *escape velocity* of the Earth, which is 11,193 m/s.

How does all of this change if we include the drag force? Let's take a look! Let us set $a = 0.2$ and $b = 0.0002$. Then the command

```
In[2]:= projMotion=proj[6.6726 10^(-11), 5.98 10^(24),
        6.37 10^6, 0.2, 0.0002, 0, 1000, 110];
        Plot[x[t] /. projMotion, {t, 0, 110}]
```

produces the plot in Figure 6.5. Note that in comparison with Figure 6.4, the maximum height of about 7 km is very small. The drag force slows down the projectile, causing it to stop at a lower altitude. Note also that as the projectile reaches the ground, the large drag force slows it down even further, accounting for the flatter slope at the bottom right portion of the curve.

Because of the drag force, we expect the terminal velocity to be larger than 11,193 m/s. In fact, changing the speed in `In[2]` to 11,500 m/s produces a curve with a maximum height of 4.8×10^7 m and a duration of 42,300 s. For the initial speed of 12,000 m/s, there will still be a maximum height; but the initial speed of 12,200 m/s will send the projectile out of the reach of the Earth's gravitational field. It turns out that the escape velocity in this case is somewhere between 12,190 m/s and 12,200 m/s.

For future reference, we record the command for solving DEs numerically in *Mathematica*:

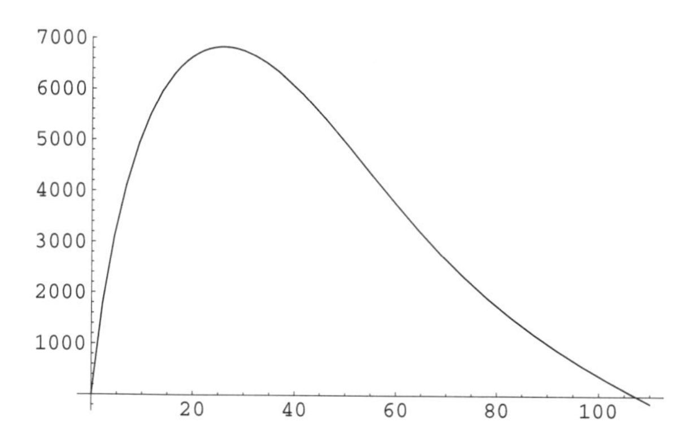

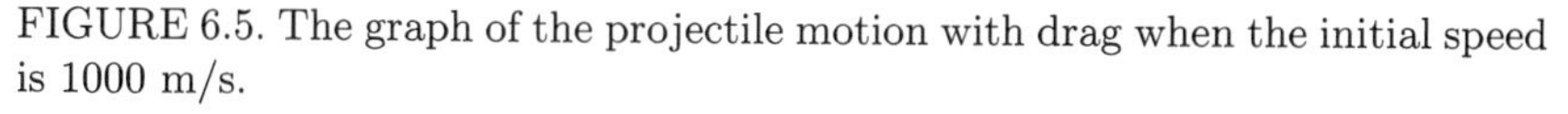
FIGURE 6.5. The graph of the projectile motion with drag when the initial speed is 1000 m/s.

`NDSolve[{DE,Con},y,{x,a,b}]`
solve `DE` with initial or boundary conditions `Con` and find y as a function of x in the range $x = a$ to $x = b$

`NDSolve[{DEs,Con},{y1,y2, ... },{x,a,b}]`
solve a system of DEs

In these commands, all `DE`s are differential *equations*, i.e., they have double equality signs. The derivatives in these equations operate on `y[x]` (not just y). Similarly, `Con` are *equations* involving linear combination of y and its derivatives, evaluated at some given points in the interval $a \leq x \leq b$.

6.3 A Two-Dimensional Projectile

Now let us bring in the other dimension. We attach a coordinate system at a point on the surface of the Earth with vertical axis labeled y and the horizontal axis labeled x as shown in Figure 6.6. It follows that the center of the Earth is at $(0, -R)$ in this coordinate system. The force of gravity can be written in vector form as

MM, p. 24

$$\mathbf{F}_{\text{grav}} = -\frac{GmM}{|\mathbf{r} - \mathbf{R}|^3}(\mathbf{r} - \mathbf{R})$$

where $\mathbf{r} = (x, y)$ and $\mathbf{R} = (0, -R)$ are the position vectors of the projectile and the Earth, respectively. In terms of the coordinates, the force becomes

$$\mathbf{F}_{\text{grav}} = -\frac{GmM}{[x^2 + (y+R)^2]^{3/2}}[x\hat{\mathbf{e}}_x + (y+R)\hat{\mathbf{e}}_y]$$

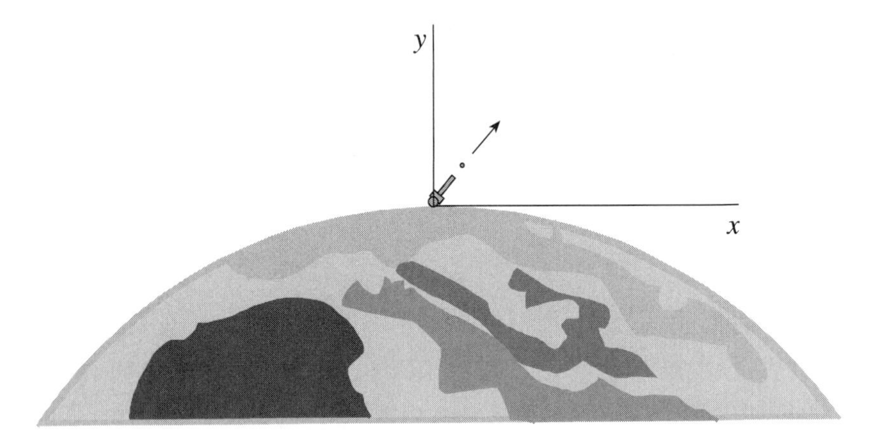

FIGURE 6.6. The coordinate system used for the two-dimensional projectile motion.

Similarly, the drag force (divided by m) can be written as

$$\mathbf{F}_{\text{drag}} = -ae^{-b\left[\sqrt{x^2+(y+R)^2}-R\right]}\mathbf{v} = -ae^{-b\left[\sqrt{x^2+(y+R)^2}-R\right]}\left(\frac{dx}{dt}\hat{\mathbf{e}}_x + \frac{dy}{dt}\hat{\mathbf{e}}_y\right)$$

The exponent of the exponential is simply the difference between the distance from the projectile to the center of the Earth and the radius of the Earth, giving the altitude above the Earth's surface. It now follows that the equations of motion we have to solve are

$$\begin{aligned}\frac{d^2x}{dt^2} &= -\frac{GM}{[x^2+(y+R)^2]^{3/2}}x - ae^{-b\left[\sqrt{x^2+(y+R)^2}-R\right]}\frac{dx}{dt} \\ \frac{d^2y}{dt^2} &= -\frac{GM}{[x^2+(y+R)^2]^{3/2}}(y+R) - ae^{-b\left[\sqrt{x^2+(y+R)^2}-R\right]}\frac{dy}{dt}\end{aligned} \tag{6.5}$$

We follow the same procedure as in the one-dimensional case. Thus, we first write the two DEs:

```
In[1]:= XeqProj2D[G_, M_, R_, a_, b_] :=
          D[x[t],{t,2}]+G M x[t]/(x[t]^2+(y[t]+R)^2)^(3/2)
          +a E^(-b (Sqrt[x[t]^2+(y[t]+R)^2]-R))D[x[t],t];
        YeqProj2D[G_, M_, R_, a_, b_] := D[y[t],{t,2}]
          +G M (y[t]+R)/(x[t]^2+(y[t]+R)^2)^(3/2)
          +a E^(-b (Sqrt[x[t]^2+(y[t]+R)^2]-R))D[y[t],t];
```

Next, we solve these equations subject to some initial conditions. This is done in a general way by typing in

```
In[2]:= proj2D[G_,M_,R_,a_,b_,x0_,y0_,vx0_,vy0_,T_] :=
          NDSolve[{XeqProj2D[G,M,R,a,b]==0,
```

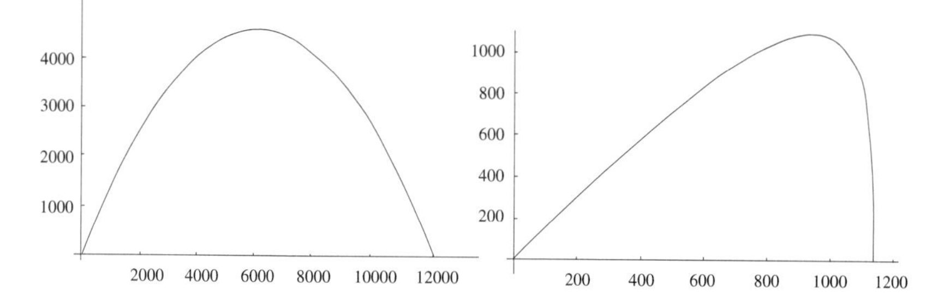

FIGURE 6.7. The projectile trajectory when there is no air drag (left), and when there is a (strong) air drag (right). At the end of the motion, the horizontal speed is practically zero, while the vertical motion—due to gravity—is still in operation.

```
YeqProj2D[G,M,R,a,b]==0,
x'[0]==vx0, y'[0]==vy0, x[0]==x0, y[0]==y0},
{x, y}, {t, 0, T}]
```

By entering appropriate parameters as arguments of proj2D, we can generate solutions for the DE with given initial (or boundary) conditions.

A convenient set of codes for this purpose is

```
In[3]:= T = 61; proj2DMotion = proj2D[6.6726 10^(-11),
        5.98 10^24, 6.37 10^6,0,0,0,0,200,300,T];
        xMotion[t_] = Part[x[t] /. proj2DMotion, 1];
        yMotion[t_] = Part[y[t] /. proj2DMotion, 1];
        ParametricPlot[{xMotion[t], yMotion[t]},
        {t, 0, T}, PlotRange -> All]
```

Let us look closely at these codes. The duration of the flight is set at the beginning by T=61. Then the DEs are solved using proj2D with some arguments, which include the initial conditions as well as the flight duration, and the solution is given the name proj2DMotion. The next two statements extract the values of x[t] and y[t] from their corresponding lists and rename them xMotion[t] and yMotion[t]. Finally, ParametricPlot plots the *actual trajectory* of the projectile. With the values given in In[3], we get the usual parabola as shown in the plot on the left in Figure 6.7.

If we want to see the effect of air drag, we change the values of a and b. For $a = 0.2$, $b = 0.0002$, $T = 36$, and the remaining parameters unchanged, we get the plot on the right in Figure 6.7. Notice how the range of the projectile is reduced from 12,000 m to less than 1200 m because of drag. Note also that the horizontal velocity is reduced considerably toward the end of the motion, so that the fall is almost vertical.

Let us now experiment with larger velocities, and type in

```
In[4]:= T = 3000; proj2DMotion = proj2D[6.6726 10^(-11),
        5.98 10^24,6.37 10^6,0,0,0,0,6000,5000,T];
```

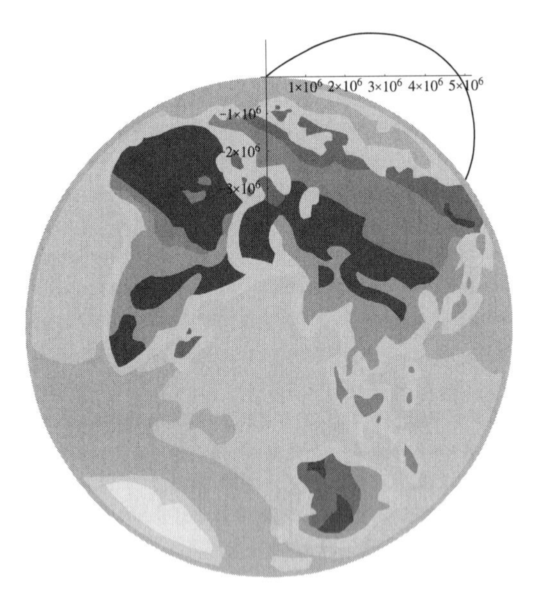

FIGURE 6.8. The projectile trajectory when there is no air drag but the velocity is fairly large.

```
xMotion[t_] = Part[x[t] /. proj2DMotion, 1];
yMotion[t_] = Part[y[t] /. proj2DMotion, 1];
ParametricPlot[{xMotion[t], yMotion[t]},
{t, 0, T}, PlotRange -> All,
AspectRatio -> Automatic]
```

We have turned off the drag force and followed the motion of the projectile for 50 min. The option `AspectRatio -> Automatic` causes the horizontal and vertical scales to be equal, so that the trajectory that we get is not deformed. For the parameters of input line 4, we get the trajectory in Figure 6.8. This Inter-Continental Ballistic Missile (ICBM) trajectory changes slightly if we introduce a drag force with $a = 0.1$ and $b = 0.0002$. However, something interesting happens when we change some of the other parameters as well.

To illustrate the point, let us increase the components of the launch velocity to $vx0 = 30$ km/s and $vy0 = 30$ km/s, the drag parameters to $a = 0.5$ and $b = 0.00002$. As we increase the flight duration, from say 500 s to about 2900 s, we see the trajectory increase in length, as we expect. However, after 2900 s we see no noticeable change in the length or shape of the trajectory. Even a flight duration of 50,000 s will not produce any change! Is the projectile stuck in mid-air? Not quite. To see what is happening, we follow the motion more closely.

We want to look at the velocity (both magnitude and direction) of the projectile at the altitudes reached toward the end of its motion. Three quantities that can help us understand this part of the motion are altitude, magnitude of the velocity, and the radial component of the velocity. We know how to calculate the altitude and the magnitude of the velocity:

$$\text{alt} = \sqrt{x^2 + (y+R)^2} - R, \qquad v \equiv |\mathbf{v}| = \sqrt{v_x^2 + v_y^2}$$

The radial component of the velocity is $\mathbf{v} \cdot \hat{\mathbf{e}}_{\text{Earth}}$, where $\hat{\mathbf{e}}_{\text{Earth}}$ is the unit vector radially outward from the center of the Earth at the location of the projectile. Since the position vector of the projectile relative to the center of the Earth is $\mathbf{r} - \mathbf{R}$, we have

MM, p. 6, Box 1.1.2

$$\hat{\mathbf{e}}_{\text{Earth}} = \frac{\mathbf{r} - \mathbf{R}}{|\mathbf{r} - \mathbf{R}|} = \frac{x\hat{\mathbf{e}}_x + (y+R)\hat{\mathbf{e}}_y}{\sqrt{x^2 + (y+R)^2}}$$

and

$$v_{\text{rad}} = \mathbf{v} \cdot \hat{\mathbf{e}}_{\text{Earth}} = \frac{xv_x + (y+R)v_y}{\sqrt{x^2 + (y+R)^2}}$$

All the information above is fed into *Mathematica* as follows:

```
In[5]:= velx[t_] := Evaluate[D[xMotion[t], t]];
        vely[t_] := Evaluate[D[yMotion[t], t]];
        velMag[t_] := Sqrt[velx[t]^2 + vely[t]^2]
        vRadial[t_] := (velx[t] xMotion[t]+vely[t]
          (yMotion[t] + 6.37 10^6))/
          Sqrt[xMotion[t]^2+(yMotion[t] + 6.37 10^6)^2];
        alt[t_] := Sqrt[xMotion[t]^2+(yMotion[t]
          + 6.37 10^6)^2] - 6.37 10^6;
        info[t_] := {t, alt[t], velMag[t], vRadial[t]}
```

The first two lines evaluate the components of the velocity; the last line outputs the information we are seeking in the form of a list.

Let us try some values for t. We rerun `In[4]` with the new parameters for $T = 5000$. Then type in

```
In[5]:= info[2800]
```

and get

$$\{2800, 186737.245, 6334.256, -4569.422\}$$

informing us that 2800 s after the launch, the projectile is at an altitude of 186,737.245 m, moving at the speed of 6334.256 m/s with a radial velocity of 4569.422 m/s *toward* the Earth. An input of `info[2850]` gives

$$\{2850, 77506., 343.27, -288.977\}$$

showing that the projectile has fallen a distance of about 109 km and slowed down considerably. With `info[3000]`, we get

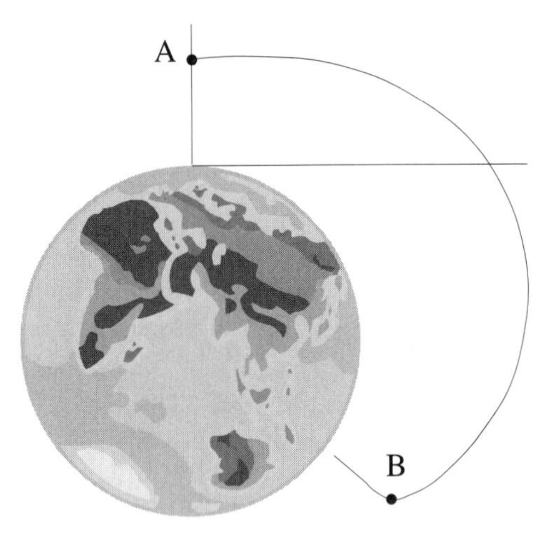

FIGURE 6.9. The projectile trajectory when there is a drag force with $a = 10$ and $b = 0.000002$. The horizontal and vertical components of velocity are 15 km/s and 1 km/s, respectively. The projectile is launched from an initial altitude of 4000 km.

$$\{3000, 63877.6, 69.8745, -69.8745\}$$

showing a further fall and a huge slowdown. The new feature of this information is that the *velocity is entirely radial*; i.e., the projectile is falling vertically down. The air drag has completely eliminated any transverse motion. All subsequent motion is entirely vertical. After some trials, we discover that $t = 4843.1$ is the approximate flight time. In fact, `info[4843.1]` yields

$$\{4843.1, 0.893226, 19.683, -19.683\}$$

suggesting that the projectile is at a height of 89.3 cm above the ground while moving at the rate of 19.7 m/s. Not a safe landing! The process of descent is very slow compared to the rest of the motion. It takes the projectile about 2900 s to go from A to B, and about 1900 s to descend vertically to Earth from B.

To see this vertical descent on the trajectory of the projectile, we need to change the parameters somewhat. Figure 6.9 illustrates the trajectory of a projectile launched from an initial altitude of 4000 km in a very thick atmosphere whose density changes slowly with height. After four and a half days, the projectile is descending vertically at a rate of less than 1.5 m/s at an altitude of approximately 240 km.

We now ignore the drag and investigate the variety of orbits obtained when the projectile is launched horizontally from a height of, say 1000 km. The minimum speed required to keep the projectile from crashing down is $\sqrt{GM/r}$—obtained by equating the gravitational acceleration $g = GM/r^2$

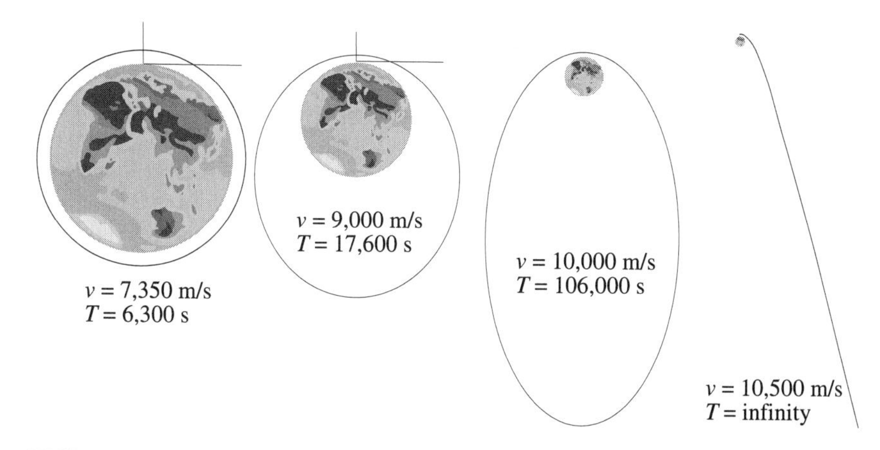

FIGURE 6.10. The different kinds of orbits for a projectile launched at a height of 1000 km with different horizontal speeds.

to the centripetal acceleration—corresponding to a circular orbit. At an altitude of 1000 km (and, therefore, $r = 7.4\times10^6$ m), this speed is 7354 m/s. The period of this orbit is 6300 s. Increasing the speed to 9000 m/s produces an elliptical orbit with major and minor axes of about 3×10^7 m and 2.5×10^7 m, respectively. The period for this orbit is about 17,600 s. Increasing the speed further to 10,000 m/s gives a longer elliptical orbit with major and minor axes of about 10^8 m and 5×10^7 m, respectively. The period for this orbit is about 106,000 s. The speed of 10,500 m/s sends the projectile out of the gravitational field of the Earth on a hyperbolic orbit. The escape velocity is 10,400 m/s at an altitude of 1000 km. All these orbits are shown in Figure 6.10.

6.4 The Two-Body Problem

The previous section discussed the motion of a projectile. In the context of what we are about to investigate, we can call the motion of a projectile a *one-body problem*: a single object under the influence of the gravitational field of an infinitely heavy immobile body (the Earth). The natural generalization of a one-body problem is a *two-body problem*, in which each of the two objects involved in the motion influences the other through a gravitational force.

Thus, in a two-body problem, we are interested in the motion of each object caused by the force of the other. Let $\mathbf{r}_1$ and $\mathbf{r}_2$ be the position vectors of the two objects relative to the origin of some Cartesian coordinate system. Then there are two vector (differential) equations to be solved:

$$\begin{aligned}\frac{d^2\mathbf{r}_1}{dt^2} &= \frac{Gm_2}{|\mathbf{r}_2-\mathbf{r}_1|^3}(\mathbf{r}_2-\mathbf{r}_1) = \frac{Gm_2}{|\mathbf{r}|^3}\mathbf{r} \\ \frac{d^2\mathbf{r}_2}{dt^2} &= \frac{Gm_1}{|\mathbf{r}_1-\mathbf{r}_2|^3}(\mathbf{r}_1-\mathbf{r}_2) = -\frac{Gm_2}{|\mathbf{r}|^3}\mathbf{r}\end{aligned} \tag{6.6}$$

where $\mathbf{r} = \mathbf{r}_2 - \mathbf{r}_1$. These constitute six second-order ODEs, corresponding to three coordinates of each of the two objects. To be able to solve these uniquely, we need 12 initial (or boundary) conditions. We can write the 6 ODEs and the 12 initial conditions manually. But there is a more elegant way, which uses some of the nice features of *Mathematica*.

We first define all the constants of our problem:

```
In[1]:= G=6.6726 10^(-11); m1=5.98 10^24; m2=7.35 10^22
```

where we have chosen the Earth (m_1) and the moon (m_2) to be our two objects. Next, we define the position vectors

```
In[2]:= r1[t_] := {x1[t], y1[t], z1[t]};
        r2[t_] := {x2[t], y2[t], z2[t]};
        r[t_] := r2[t] - r1[t];
```

We now want to write the DEs of Equation (6.6) with all the terms moved to the left-hand side. The resulting left-hand sides are

```
In[3]:= eq1=D[r1[t],{t,2}]-G m2 r[t]/(r[t].r[t])^(3/2);
        eq2=D[r2[t],{t,2}]+G m1 r[t]/(r[t].r[t])^(3/2);
```

Notice that by differentiating a vector, we differentiate all its components. We have also used the dot product to evaluate the denominator.

using **Thread**

Our next task is to turn these into a set of six DEs. We use the command `Thread` to do that. Basically `Thread` applies a function or procedure to the elements of a list. For example, `Thread[f[r1[t],r2[t]]]` produces

```
{f[x1[t], x2[t]], f[y1[t], y2[t]], f[z1[t], z2[t]]}
```

Thus, we implement the boundary conditions (BCs) as

```
In[4]:= BC11 = Thread[r1[0] == {0, 0, 0}];
        BC12 = Thread[Evaluate[r1'[0]] == {0, 0, 0}];
        BC21 = Thread[r2[0] == {3.85 10^8, 0, 0}];
        BC22 = Thread[Evaluate[r2'[0]] == {0, 1022, 0}];
```

The first line is equivalent to

```
BC11 ={x1[0]==0, y1[0]==0, z1[0]==0}
```

and places the Earth at the origin. The second line fixes the initial velocity of the Earth to zero. The third line places the moon on the x-axis 3.85×10^8 m away from the origin (or Earth). Finally, the last line gives the moon an initial velocity of 1022 m/s in the positive y-direction.

We can thread the DEs as well. For example, `Thread[eq1==0]` produces a *list* of DEs involving the components of $\mathbf{r}_1$; similarly with $\mathbf{r}_2$. This way, we create a number of separate lists for DEs and boundary conditions. To solve these numerically, we have to `Join` them. So, our next command will be

using **Join**

```
In[5]:= listEq=Join[Thread[eq1==0],Thread[eq2==0],
          BC11, BC12, BC21, BC22];
```

which creates a single list consisting of 6 DEs and 12 BCs. We can now tell *Mathematica* to solve them:

```
In[6]:= TwoBodies = NDSolve[listEq, Join[r1[t], r2[t]],
          {t, 0, 5 10^6}];
        Earth[t_] = r1[t] /. TwoBodies;
        moon[t_] = r2[t] /. TwoBodies;
```

The arguments of `NDSolve` are a list of DEs and BCs called `listEq`, a list of independent variables consisting of the three components of $\mathbf{r}_1$ and three components of $\mathbf{r}_2$ combined using `Join`, and a last list giving the independent variable and its range—chosen to be about two revolutions of the moon. The last two lines in `In[6]` define the Earth position `Earth[t]` and the moon position `moon[t]` as the solutions of the system of DEs.

Now we can plot the trajectories of the moon and Earth. Typing in

```
In[7]:= ParametricPlot3D[{moon[t][[1,1]],moon[t][[1,2]],
        moon[t][[1, 3]]}, {t, 0, 5 10^6}, Ticks -> False]
```

produces the curve shown on the left of Figure 6.11.

The brackets after `moon[t]` in input 7 pick entries of that list. To demonstrate this, type in `moon[1000]` and get

$$\{\{3.84999 \times 10^8, 1.022 \times 10^6, 0.\}\}$$

which is a list of one list with three entries, or a *two-level* list. The first level is obtained by typing `moon[1000][[1]]`, which puts out

$$\{3.84999 \times 10^8, 1.022 \times 10^6, 0.\}$$

i.e., the only element—which happens to be a list itself—of the outer list. Each element of this inner list can be selected by entering a second number after 1 separated by a comma. Thus, `moon[1000][[1,1]]` yields 3.84999×10^8, etc.

Figure 6.11 shows some interesting results. First of all, the entire motion of the two bodies takes place in the xy-plane. This is to be expected, because

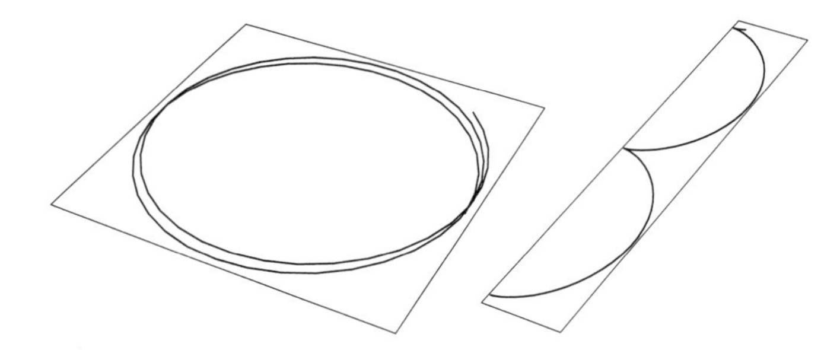

FIGURE 6.11. The orbit of moon (left) and Earth (right). The departure from elliptical shape is due to the motion of the center of mass of the system.

the entire initial conditions occur in the same plane; and since the forces on the two bodies have no components perpendicular to this plane, the bodies cannot accelerate out of the plane. Second, the orbit (left of the figure) of the moon is "moving" along the y-axis, so that after each revolution, the moon starts on a new orbit displaced from the previous one by a small amount. The reason is that the *center of mass of the system is moving in the y-direction.* When we start the moon in the y-direction with a speed of $v_{20} = 1022$ m/s, the center of mass picks up a speed of

$$v_{\text{cm}} = \frac{m_2 v_{20}}{m_1 + m_2} = \frac{7.35 \times 10^{22} \times 1022}{5.98 \times 10^{24} + 7.35 \times 10^{22}} = 12.4 \text{ m/s}$$

which in the course of a revolution (about 27 days) adds up to about 2.9×10^7 m, a noticeable displacement. Third, the Earth orbit (on the right of the figure), which is a circle about the center of mass (see below), becomes a *cycloid*—the curve traced by a point on the rim of a moving wheel. Indeed, the Earth is a point on the wheel (its orbit) that moves uniformly due to the uniform motion of the center of mass.

Let us now change the initial conditions so that the center of mass is at the origin and has zero momentum initially (and, therefore, for all times). This can be done by positioning the Earth at $x = -[m_2/(m_1 + m_2)]3.85 \times 10^8$, the moon at $x = -[m_1/(m_1 + m_2)]3.85 \times 10^8$, and giving the Earth a momentum equal in magnitude to the initial momentum of the moon, but opposite in direction. The speed corresponding to this momentum is $(m_2/m_1)v_{20}$. So, we change input line 4 to

```
In[8]:= BC11=Thread[r1[0]=={-(m2/(m1+m2)) 3.85 10^8,0,0}];
        BC12 = Thread[Evaluate[r1'[0]] ==
          {0, -(m2/m1) 1022, 0}];
        BC21 = Thread[r2[0] ==
          {(m1/(m1+m2))3.85 10^8, 0, 0}];
        BC22 = Thread[Evaluate[r2'[0]] == {0, 1022, 0}];
```

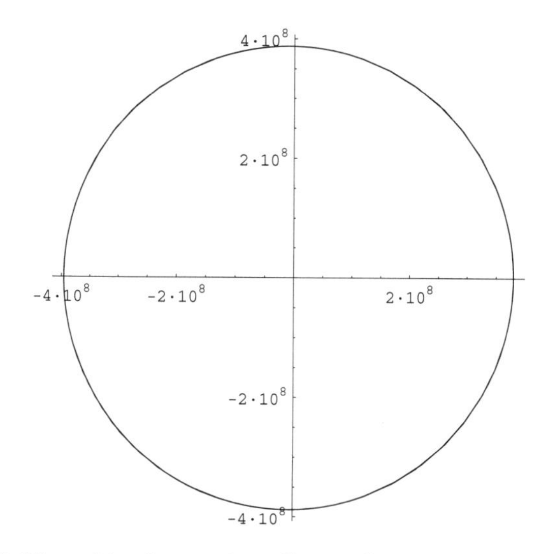

FIGURE 6.12. The orbit of moon is really an ellipse whose foci are very close.

and rerun input lines 5, 6, and 7. Once again, the motion of both objects will be confined to the xy-plane. So, instead of a three-dimensional plot, we make an ordinary plot of the trajectory of the Earth and the moon. For example, for the moon, we type in

```
In[9]:= ParametricPlot[{moon[t][[1,1]],moon[t][[1,2]]}
         {t, 0, 5 10^6}]
```

and obtain the plot in Figure 6.12. Since the center of mass is at the origin this time, the moon retraces its orbit in its second revolution (the upper limit of t in `In[9]` is over twice the moon's period).

Although it is not apparent from the plot, the moon's orbit is an ellipse with the center of mass (the origin) at one of its foci. In fact, by comparing the location of the moon at $t = 0$ and at $t = T/2$, where T is the period of the moon, one can find the distance between the two foci and show that it is not zero. The details are left as an exercise for the reader. A careful examination of Figure 6.12 indeed reveals that the origin of the figure is not the center of the "circle."

One can also plot the orbit of the Earth. That orbit is also an ellipse with very small eccentricity (a circle has zero eccentricity). It is interesting to note that the center of mass of the Earth–moon system is inside the Earth. So, when we talk about the orbit of the Earth, we mean the motion of the center of the Earth about the center of mass of the system.

In all the discussion above, the moon and Earth have been confined to the xy-plane, because the initial conditions were confined to that plane. If we give one of the objects an initial velocity along the z-axis, the motion

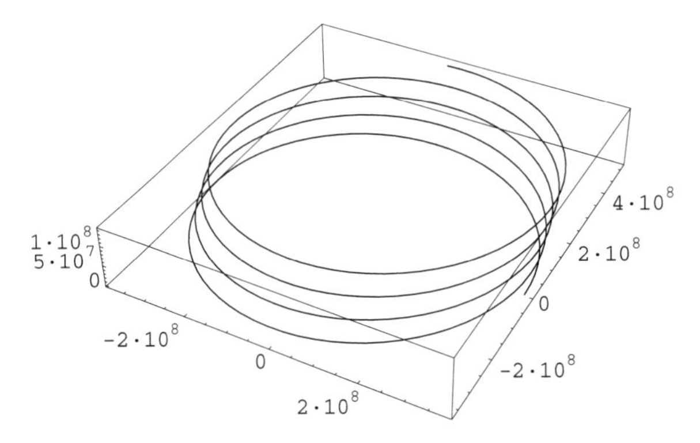

FIGURE 6.13. The orbit of moon when Earth is given an initial speed along the z-direction.

will no longer take place solely in a plane. To see the effect of such an initial condition, type in

```
In[10]:= BC11=Thread[r1[0]=={-(m2/(m1+m2)) 3.85 10^8,0,0}];
         BC12=Thread[Evaluate[r1'[0]]=={0, 0, 15}];
         BC21=Thread[r2[0]=={(m1/(m1+m2))3.85 10^8,0,0}];
         BC22=Thread[Evaluate[r2'[0]]=={0,1022,0}];
```

giving the Earth an initial velocity of 15 m/s along the z-axis. Now run input line 5 to set up the DEs, and 6 for 10^7 seconds to obtain the trajectories. A three-dimensional parametric plot of the moon's trajectory is shown in Figure 6.13. Notice how the entire orbit is displaced along the y- and z-axes. This is, of course, due to the uniform motion of the center of mass, which, due to the initial velocities, has a y- as well as a z-component.

6.4.1 Precession of the Perihelion of Mercury

One of the earliest triumphant tests of the general theory of relativity (GTR) was the precession of the orbits of planets. This precession is so small that it is observable only for Mercury, the closest planet to the Sun, and, therefore, in its strongest gravitational field.

Over many decades astronomers had noted that the orbit of Mercury precessed about 532 arcseconds per century. Tidal perturbations due to other planets accounted for 489 arcseconds per century, leaving 43 arcseconds per century unaccounted for. When Einstein's GTR was applied to the motion of planets, it was noted that it too causes a precession of the orbits, which only in the case of Mercury was large enough to be measurable. In fact, GTR predicts—to within 1% accuracy—the same amount of precession that was unaccounted for classically.

To the first nontrivial approximation, Einstein's GTR changes the force of gravity by adding a force that varies as the inverse fourth power of distance. To be precise, the force of gravity due to a (large) mass m_1 on a (small) mass m_2 including Einstein's correction is

$$\mathbf{F}_{12} = -\frac{Gm_1m_2}{|\mathbf{r}_2 - \mathbf{r}_1|^3}(\mathbf{r}_2 - \mathbf{r}_1) - \frac{3L_2^2Gm_1m_2}{c^2|\mathbf{r}_2 - \mathbf{r}_1|^5}(\mathbf{r}_2 - \mathbf{r}_1)$$

where L_2 is the angular momentum of m_2 and c is the speed of light. We will lump all the constants in the second term into one constant and call it a.

Instead of the Earth–moon system discussed above, let us consider the Sun–Mercury system. Since we are impatient and don't want to wait a hundred years to see only 43 arcseconds, we change some of the actual parameters to expedite the precession. Two major changes are important for our consideration. First, the (initial) orbital speed has to be increased from the actual 47.9 km/s to 60 km/s to change the almost-circular orbit of Mercury to a well-defined ellipse, because it is very hard to see a precession if the orbit is almost circular. Second, to make the precession larger, we change the actual value of a, which is approximately 10^{34}, to 10^{40}. Thus, we use the following parameters:

```
In[1]:= G=6.6726 10^(-11);m1=2 10^30;m2=3.3 10^23;a=10^40
```

which, except for a, are the actual data for the Sun and Mercury.

The definitions of $\mathbf{r}_1$, $\mathbf{r}_2$, and $\mathbf{r}$ are as before. Thus input line 2 on page 192 will not change. However, input line 3 becomes

```
In[3]:= eq1=D[r1[t],{t,2}]-G m2 r[t]/(r[t].r[t])^(3/2);
        eq2=D[r2[t],{t,2}]+G m1 r[t]/(r[t].r[t])^(3/2)
          + a r[t]/(r[t].r[t])^(5/2);
```

For boundary conditions, we change input line 8 to

```
In[8]:= BC11=Thread[r1[0]=={-(m2/(m1+m2)) 5.79 10^(10),
          0,0}];
        BC12 = Thread[Evaluate[r1'[0]] ==
          {0, -(m2/m1) 60000, 0}];
        BC21 = Thread[r2[0] ==
          {(m1/(m1+m2)) 5.79 10^(10), 0, 0}];
        BC22 = Thread[Evaluate[r2'[0]] == {0, 60000, 0}];
```

with a larger initial speed. The heart of the calculation—input line 6—is also changed accordingly:

```
In[6]:= TwoBodies = NDSolve[listEq, Join[r1[t], r2[t]],
          {t, 0, 3 10^8}];
        Sun[t_] = r1[t] /. TwoBodies;
        Mercury[t_] = r2[t] /. TwoBodies;
```

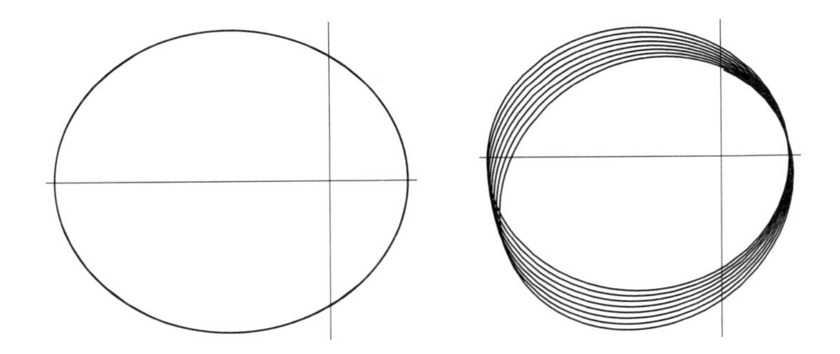

FIGURE 6.14. Left: The orbit of Mercury without relativistic correction. Right: The precession of orbit of Mercury due to the relativistic correction.

A parametric plot of Mercury's orbit for $t = 2 \times 10^8$ s produces the plot on the right of Figure 6.14. With $a = 0$, i.e., without relativistic correction, we get the plot on the left. It is clear that the perihelion of the orbit of Mercury precesses because of the relativistic correction.

6.5 The Three-Body Problem

All the nonrelativistic calculations in the previous section can be done analytically. In fact, historically that calculation was the first application of Newton's laws of motion to a "solar system" consisting of a Sun and a single planet. This *Kepler problem* was the first of a long series of triumphs of the classical laws of motion. However, when one tries to apply the laws of motion to a real solar system with several planets of different masses, one encounters difficulties. In fact, adding a single new body to a two-body system renders it impossible to solve analytically. But with *Mathematica* at our disposal, we can resort to a numerical solution.

6.5.1 *Massive Star and Two Planets*

The generalization of the two-body problem to the three-body problem is straightforward. The first statement is

```
In[1]:= G=6.6726 10^(-11); m1=2 10^30; m2=5.98 10^24;
        m3=3.3 10^23; x10=0; v10=0; x20=1.5 10^(11);
        v20 = 29800; x30 = 5.79 10^(10); v30 = 47900;
```

where we have included the nonzero initial position coordinates and velocities. All data given here are actually those of the Sun, Earth, and Mercury. We are not interested in the motion perpendicular to the xy-plane (i.e., we fix the initial conditions so that no such motion develops). Thus, all our vectors are two-dimensional, and the second statement is

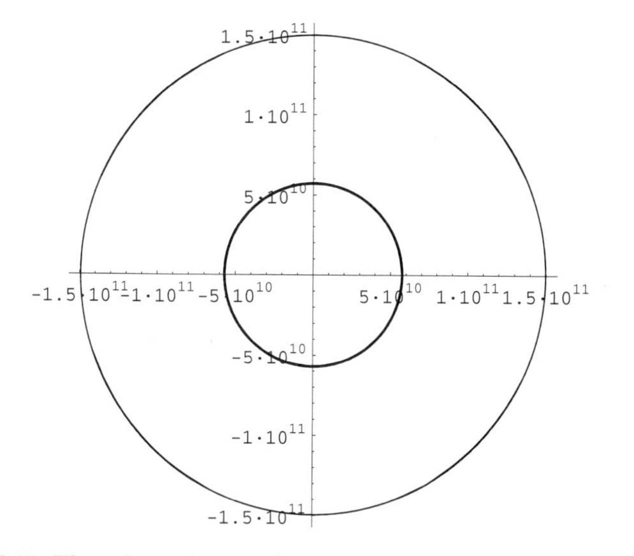

FIGURE 6.15. The orbits of Earth (larger circle) and Mercury retraced for 10 years (10 revolutions of Earth).

```
In[2]:= r1[t_]:={x1[t],y1[t]}; r2[t_]:={x2[t],y2[t]};
        r3[t_]:={x3[t],y3[t]}; r12[t_]:=r2[t]-r1[t];
        r13[t_]:=r3[t]-r1[t]; r23[t_]:=r3[t]-r2[t];
```

The left-hand sides of the differential equations are

```
In[3]:= eq1=D[r1[t],{t,2}]
          -G m2 r12[t]/(r12[t].r12[t])^(3/2)
          -G m3 r13[t]/(r13[t].r13[t])^(3/2)
        eq2=D[r2[t],{t,2}]
          +G m1 r12[t]/(r12[t].r12[t])^(3/2)
          -G m3 r23[t]/(r23[t].r23[t])^(3/2)
        eq3=D[r3[t],{t,2}]
          +G m1 r13[t]/(r13[t].r13[t])^(3/2)
          +G m2 r23[t]/(r23[t].r23[t])^(3/2)
```

and if we line up the Sun, Earth, and Mercury along the x-axis and give them initial velocities along the y-axis, then the BCs will look like this:

```
In[4]:= BC11 = Thread[r1[0] == {x10, 0}];
        BC12 = Thread[Evaluate[r1'[0]] == {0, v10}];
        BC21 = Thread[r2[0] == {x20, 0}];
        BC22 = Thread[Evaluate[r2'[0]] == {0, v20}];
        BC31 = Thread[r3[0] == {x30, 0}];
        BC32 = Thread[Evaluate[r3'[0]] == {0, v30}];
```

The system of DEs is obtained by the use of `Thread` and `Join`:

```
In[5]:= listEq=Join[Thread[eq1==0],Thread[eq2==0],
          Thread[eq3==0],BC11,BC12,BC21,BC22,BC31,BC32];
```

Finally, we solve these equations and extract the solutions for the Sun, Earth, and Mercury:

```
In[6]:= ThreeBodies = NDSolve[listEq, Join[r1[t], r2[t],
          r3[t]],{t, 0, 5 10^6}];
        Sun[t_] = r1[t] /. ThreeBodies;
        Earth[t_] = r2[t] /. ThreeBodies;
        Mercury[t_] = r3[t] /. ThreeBodies;
```

With the solution found, we can look at some plots. For instance,

```
In[7]:= ParametricPlot[
          {{Mercury[t][[1,1]], Mercury[t][[1,2]]},
          {Earth[t][[1,1]], Earth[t][[1,2]]}},
          {t, 0, 3.2 10^8}, AspectRatio->Automatic]
```

gives the orbits of Figure 6.15. It shows that—at least for 3.2×10^8 s, or 10 Earth years—the orbits are retraced.

What does the orbit of Mercury look like as seen from Earth? To answer that, we need to find the position vector of Mercury relative to Earth. But this is simply $\mathbf{r}_3 - \mathbf{r}_2$. Therefore, the command should be of the form

```
In[8]:= ParametricPlot[{Mercury[t][[1,1]]-Earth[t][[1,1]],
          Mercury[t][[1,2]]-Earth[t][[1,2]]},
          {t,0,6.4 10^7}, Ticks -> False,
          AspectRatio -> Automatic]
```

the retrograde motion of Mercury relative to Earth

which produces the trajectory of Figure 6.16. This trajectory reveals a feature known to the Hellenistic astronomers and for the explanation of which they introduced the idea of the *epicycle*: Mercury loops backward during some parts of its motion, while it also gets closer to Earth. This part of Mercury's motion is called the *retrograde* motion.

6.5.2 *Light Star and Two Planets*

The enormous mass of the Sun overshadows the other gravitational fields, and, therefore, controls the motion of each planet as if the other planet were not there. By changing the masses of the three bodies and their initial positions and velocities, we can create very erratic trajectories. To reduce the dominance of the Sun, we make Earth and Mercury more massive. Consider the following parameters:

```
In[9]:= G=6.6726 10^(-11); m1=2 10^30; m2= 10^29;
        m3=6 10^28; x10=0; v10=0; x20=1.5 10^(11);
        v20 = 30000; x30 = 8 10^(10); v30 = 45000;
```

FIGURE 6.16. The trajectory of Mercury as seen from Earth (at the origin). Note the retrograde loops at which Mercury gets closest to Earth.

The masses of Earth and Mercury are now, respectively, 5% and 3% of the Sun's mass, and Mercury is initially closer to Earth than before. Feeding these parameters to the system of DEs with the time interval between 0 and 5×10^7 s and plotting the two trajectories on the same graph for the same time interval produce the left diagram in Figure 6.17, which is in complete contrast to the nice periodic trajectories of Figure 6.15.

The complicated trajectories of Figure 6.17 herald the dangerous possibility of a close encounter of the two planets. To examine this possibility, we once again look at the trajectory of Mercury relative to Earth by typing in

```
In[10]:= ParametricPlot[{Mercury[t][[1,1]]-Earth[t][[1,1]],
          Mercury[t][[1,2]]-Earth[t][[1,2]]},{t,0,5 10^7},
          Ticks -> False, AspectRatio -> Automatic]
```

corresponding to a time span of 5×10^7, or about a year and a half. If the trajectory passes through the origin, where the Earth is located, we have a collision. The plot on the right in Figure 6.17 indicates a possible collision. However, we have not given the system sufficient time to run its course. So, we increase the upper limit of the time interval to 6×10^7 s in the argument of `NDSolve`. *Mathematica* will give the following error message:

```
NDSolve::"ndsz": "At t == 5.204117984913484×10^7, step size
  is effectively zero; singularity suspected."
```

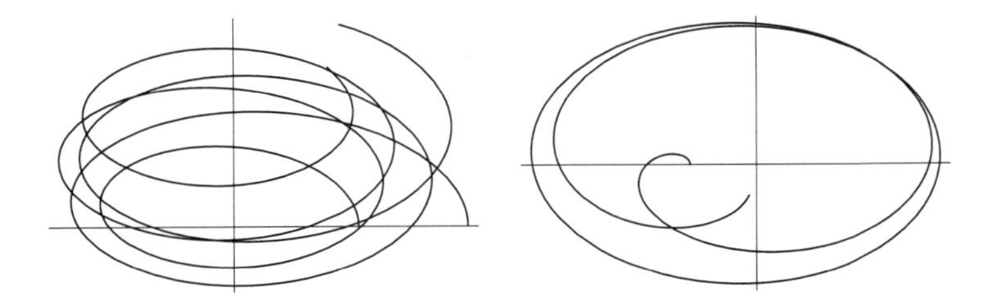

FIGURE 6.17. Left: The orbits of heavy Mercury and Earth around a light Sun. Right: The trajectory of Mercury as seen from Earth (origin).

This should give us a clue about the time of close encounter! Thus, we change the upper limit of the time interval to $5.204117984913484 \times 10^7$ s in the argument of `NDSolve`.

How close do the planets get at this "singularity" time? To find out, write the distance between them as a function of time:

```
In[11]:= dist[t_]:=Sqrt[
          (Mercury[t][[1, 1]] - Earth[t][[1, 1]])^2
          + (Mercury[t][[1, 2]] - Earth[t][[1, 2]])^2]
```

Then

```
In[12]:= dist[5.204117984913484 10^7]
```

yields 476,179. The centers of the two planets have come to within 476 km of one another! This indicates a collision. We have to emphasize that this is a strictly three-body phenomenon. Two-body collisions occur if a planet has zero or very small initial velocity component perpendicular to the line joining its center to the Sun. Once this transverse component of velocity is large enough, there will never be a collision. The transverse velocities in input line 9 are large enough to set each planet *separately* on a stable orbit. However, once the three bodies interact with one another, those same velocities put two of them on a collision course in less than two years. For a detailed study of the three-body problem using *Mathematica*, see [Gass 98, pp. 91–117].

6.6 Nonlinear Differential Equations

A variety of techniques, in particular, the method of infinite power series, could solve almost all *linear* DEs of physical interest. However, some very fundamental questions such as the stability of the solar system led to DEs that were not linear, and for such DEs no analytic (including series representation) solution existed. In the 1890s, Henri Poincaré, the great French mathematician, took upon himself the task of gleaning as much information from the DEs describing the whole solar system as possible. The result was the invention of one of the most powerful branches of mathematics (topology) and the realization that the *qualitative* analysis of (nonlinear) DEs could be very useful.

One of the discoveries made by Poincaré, which much later became the cornerstone of many developments, was that *unlike the linear DEs, nonlinear DEs may be very sensitive to the initial conditions.* In other words, if a nonlinear system starts from some initial conditions and develops into a certain final configuration, then starting it with *slightly* different initial conditions may cause the system to develop into a final configuration completely different from the first one. This is in complete contrast to the linear DEs, where two nearby initial conditions lead to nearby final configurations.

In general, the initial conditions are not known with infinite accuracy. Therefore, the final states of a nonlinear dynamical system may exhibit an indeterministic behavior resulting from the initial (small) uncertainties. This is what has come to be known as **chaos**.

chaos due to uncertainty in initial conditions

Although analytic solutions are known only for a handful of nonlinear DEs, the preponderance of computational tools has made the numerical solution of these DEs conveniently possible. With personal computers becoming a household item, and fast powerful calculational software such as *Mathematica* being available, an explosion of interest has been revived in studying nonlinear differential equations. In this section we examine one example of nonlinear DEs.

As a paradigm of a nonlinear dynamical system, we shall study the motion of a harmonically driven dissipative pendulum whose angle of oscillation is not necessarily small. The equation of motion of such a pendulum, coming directly from the second law of motion, is

$$m\frac{d^2x}{dt^2} = F_0\cos(\Omega t) - b\frac{dx}{dt} - mg\sin\theta, \tag{6.7}$$

where x is the length (as measured from the equilibrium position) of the arc of the circle on which mass m moves (see Figure 6.18).

The first term on the RHS of Equation (6.7) is the harmonic driving force with angular frequency Ω, the second is the dissipative (friction, drag, etc.) force, and the last is the gravitational force in the direction of motion. The minus signs appear because the corresponding forces oppose the motion. Since the pendulum is confined to a circle, x and θ are related via $x = l\theta$, and we obtain

$$ml\frac{d^2\theta}{dt^2} = F_0\cos(\Omega t) - bl\frac{d\theta}{dt} - mg\sin\theta$$

Let us change t to $t = \tau\sqrt{l/g}$, where τ is a dimensionless parameter measuring time in units of $T/(2\pi)$ with $T = 2\pi\sqrt{l/g}$ being the period of the

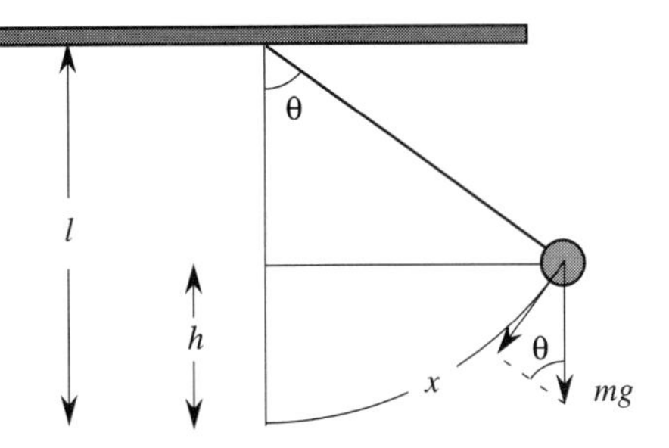

FIGURE 6.18. The displacement x and the gravitational force acting on the pendulum.

small-angle pendulum. Then, with the dimensionless constants

$$\gamma \equiv \frac{b}{m}\sqrt{\frac{l}{g}}, \qquad \phi_0 \equiv \frac{F_0}{mg}, \qquad \omega_D \equiv \Omega\sqrt{\frac{l}{g}}$$

the DE of motion becomes

$$\ddot{\theta} + \gamma\dot{\theta} + \sin\theta = \phi_0 \cos(\omega_D t) \tag{6.8}$$

where now t is the "dimensionless" time, and the dot indicates differentiation with respect to this t.

phase-space diagram

One of the devices that facilitates our understanding of nonlinear dynamical systems is the **phase-space diagram**. The phase space of a dynamical system is a Cartesian multidimensional space whose axes consist of positions and momenta of the particles in the system. Instead of momenta the velocities of particles are mostly used. Thus a single particle confined to one dimension (such as a particle in free fall, a mass attached to a spring, or a pendulum) has a two-dimensional phase space corresponding to the particle's position and speed. Two particles moving in a single dimension have a four-dimensional phase space corresponding to two positions and two speeds. A single particle moving in a plane also has a four-dimensional phase space because two coordinates are needed to determine the position of the particle, and two components to determine its velocity, and a system of N particles in space has a $6N$-dimensional phase space.

phase-space trajectory

A **phase-space trajectory** of a dynamical system is a curve in its phase space corresponding to a possible motion of the system. If we can solve the equations of motion of a dynamical system, we can express all its position and velocity variables as a function of time, constituting a parametric equation of a curve in phase space. This curve is the trajectory of the dynamical system.

Let us consider the simplest pendulum problem in which there is no driving force, the dissipative effects are turned off, and the angle of oscillation is small. Then (6.8) reduces to the *linear* DE $\ddot{\theta}+\theta = 0$, whose most general solution is $\theta = A\cos(t+\alpha)$ so that

$$\theta = A\cos(t+\alpha)$$
$$\omega \equiv \dot{\theta} \equiv \frac{d\theta}{dt} = -A\sin(t+\alpha) \tag{6.9}$$

This is a one-dimensional system (there is only one coordinate, θ) with a two-dimensional phase space. Equation (6.9) is the parametric equation of a circle of radius A in the $\theta\omega$-plane. Because A is arbitrary (it is, however, determined by initial conditions), there are (infinitely) many trajectories for this system.

Let us now make the system a little more complicated by introducing a dissipative force, still keeping the angle small. The DE (still linear) is now

$$\ddot{\theta} + \gamma\dot{\theta} + \theta = 0$$

and the general solution for the damped oscillatory case is

$$\theta(t) = Ae^{-\gamma t/2}\cos(\omega_0 t + \alpha) \quad \text{where} \quad \omega_0 \equiv \frac{\sqrt{4-\gamma^2}}{2}$$

with

$$\omega = \dot{\theta} = -Ae^{-\gamma t/2}\left\{\frac{\gamma}{2}\cos(\omega_0 t + \alpha) + \omega_0\sin(\omega_0 t + \alpha)\right\}$$

The phase-space trajectories of this system can be plotted using *Mathematica*'s `ParametricPlot`. Two such trajectories for $A = 1$ and $A = 2$ (but the same γ of 0.5) are shown in Figure 6.19.

A new feature of this system is that regardless of where the trajectory starts at $t = 0$, it will terminate at the origin. The analytic reason for this is of course the exponential factor in front of both coordinates which will cause their decay to zero after a long (compared to $1/\gamma$) time. It seems that the origin "attracts" *all* trajectories and for this reason is called an **attractor**.

attractor

We now turn to the *numerical* solution of the motion of the pendulum. To be able to have access to the derivative of the angle, we turn the second-order DE of the driven pendulum into a set of first-order DEs. First we rewrite the DE describing a general pendulum [see Equation (6.8)] as

$$\ddot{\theta} + \gamma\dot{\theta} + \sin\theta = \phi_0\cos\alpha$$

where α is simply $\omega_D t$. Then turn this equation into the following entirely equivalent set of three first-order *nonlinear* DEs:

$$\dot{\theta} = \omega, \qquad \dot{\omega} = -\gamma\omega - \sin\theta + \phi_0\cos\alpha, \qquad \dot{\alpha} = \omega_D \tag{6.10}$$

The two-dimensional (θ, ω) phase space has turned into a three-dimensional (θ, ω, α) phase space. The reason behind introducing the new variable α

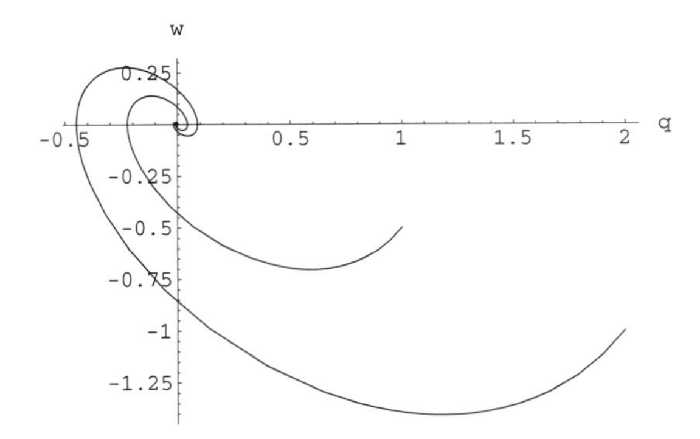

FIGURE 6.19. The phase-space trajectories of a damped pendulum undergoing small-angle oscillations with no driving force. Different spirals correspond to different initial conditions.

is to avoid explicit appearance of the independent variable t. It turns out that these **autonomous** systems—systems that have no explicit time dependence—are more manageable than DEs with explicit time dependence.

A brief *Mathematica* code for solving Equation (6.10) numerically is

```
In[1]:= pend[θ0_, ω0_, γ_, ϕ0_, ωD_, T_] :=
        NDSolve[{θ'[t]==ω[t],
          ω'[t]==-γω[t]-Sin[θ[t]]+ϕ0 Cos[α[t]],
          α'[t]==ωD,
          θ[0]==θ0, ω[0]==ω0, α[0] == 0},
          {θ, ω, α}, {t, 0, T}];
```

where we have used the *Mathematica* palette containing the Greek alphabet.

Just as in the linear case, it is instructive to ignore the damping and driving forces first. We set γ and ϕ_0 equal to zero in Equation (6.10) and solve the set of DEs numerically. For small angles, we expect a simple harmonic motion (SHM) whose phase-space diagram is a circle. We are interested in the phase-space diagrams when the maximum angular displacements are large. With the initial angular velocity set at zero, the pendulum will exhibit a periodic behavior represented by closed loops in the phase space. Figure 6.20 shows four such closed loops corresponding—from small to large loops—to the initial angular displacement of $\pi/5$, $\pi/2$, $2\pi/3$, and (almost) π. These loops are produced by typing in

```
In[2]:= sol1 = pend[Pi/5, 0, 0, 0, 0, 100];
        sol2 = pend[Pi/2, 0, 0, 0, 0, 100];
        sol3 = pend[2 Pi/3, 0, 0, 0, 0, 100];
        sol4 = pend[Pi/1.0001, 0, 0, 0, 0, 100];
```

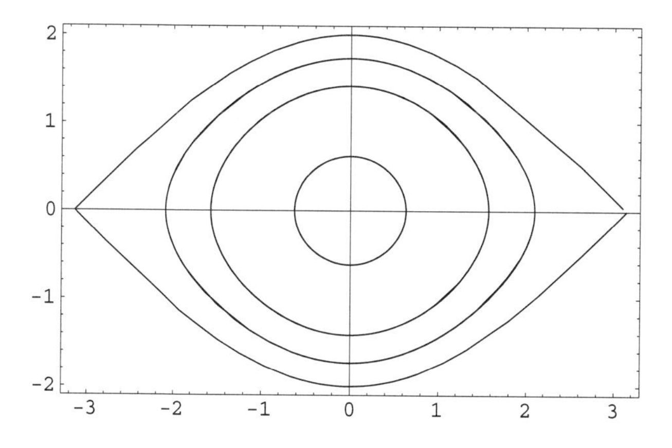

FIGURE 6.20. Phase-space diagrams for a pendulum corresponding to different values of initial (maximum) displacement angles (horizontal axis). The initial angular speed is zero for all diagrams.

and

```
In[3]:= ParametricPlot[Evaluate[{{θ[t], ω[t]}/.sol1,
          {θ[t], ω[t]}/.sol2, {θ[t], ω[t]}/.sol3,
          {θ[t], ω[t]}/.sol4}], {t, 0, 30},
          PlotRange -> All, Frame -> True]
```

They represent *oscillations only*: the angular displacement is bounded between a minimum and a maximum value determined by $\theta(0)$. The closed loops are characterized by the fact that the angular speed vanishes at maximum (or minimum) θ, allowing the pendulum to start moving in the opposite direction. Note that the time parameter in the plot of `In[3]` goes up to 30 rather than 100. This is because only 30 units of time are required to complete all the closed loops. In fact, the smaller loops need less time to close on themselves; only the largest loop needs a minimum of 30 units to get completed. The smaller loops retrace themselves during the extra times.

Suppose now that we set $\theta(0) = -\pi$ and $\omega(0) = 1$, corresponding to raising the pendulum all the way up until it is above its pivot, and giving it an initial speed. The angular displacement is unbounded: it keeps increasing for all times. Physically, this corresponds to forcing the pendulum to "go over the hill" at the top by providing it an initial angular velocity. If the pendulum is pushed over this hill once, it will continue doing it forever because there is no damping force. The rotations are characterized by a nonzero angular velocity at $\theta = \pm\pi$. This is clearly shown in Figure 6.21, produced by typing in

```
In[4]:= sol5 = pend[-Pi, 1, 0, 0, 0, 100];
```

and

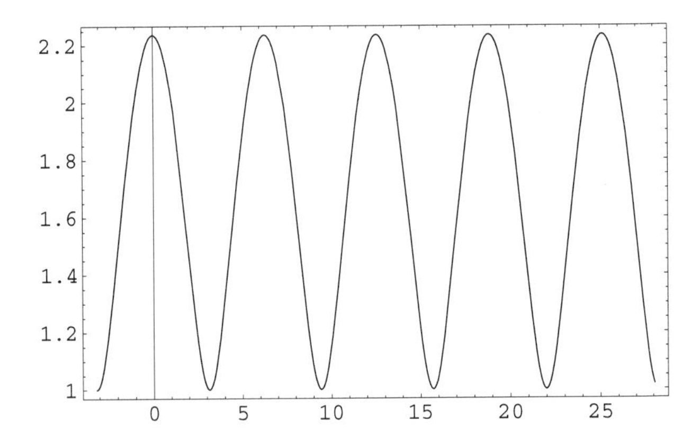

FIGURE 6.21. Phase-space diagrams for a pendulum corresponding to a maximum displacement angles of $-\pi$ and an initial angular speed of $\omega = 1$.

```
In[5]:= ParametricPlot[Evaluate[{θ[t], ω[t]}/.sol5],
          {t, 0, 20}, PlotRange -> All, Frame -> True]
```

Note that the horizontal axis θ is not bounded: if we increase T beyond 20, θ will also increase. However, the angular speed ω (the vertical axis) will remain bounded because of the conservation of energy.

If the damping force is turned on (still without any driving force), the trajectories will spiral into the origin of the phase space as in the case of the linear (small-angle) pendulum. However, the interesting motion of a pendulum begins when we turn on a driving force regardless of whether or not the dissipative effect is present. Nevertheless, let us place the pendulum in an environment in which $\gamma = 0.3$. Now drive this pendulum with a (harmonic) force of amplitude $\phi_0 = 0.5$ and angular frequency $\omega_D = 1$. For $\theta_0 = \pi$ and $\omega_0 = 0$, Equation (6.10) will then give a solution that has a transient motion lasting until $t \approx 32$. From $t = 32$ onward, the system traverses a closed orbit in the phase diagram as shown in Figure 6.22. This orbit is an *attractor* in the same sense as a point is an attractor for a dissipative nondriven pendulum. An attractor such as the one exhibited in Figure 6.22 is called a **limit cycle**.

limit cycle

As we increase the control parameter ϕ_0, the phase-space trajectories go through a series of periodic limit cycles until they finally become completely aperiodic: chaos sets in. Figure 6.23 shows four trajectories whose common initial angular displacement θ_0, initial angular velocity ω_0, damping factor γ, and drive frequency ω_D are, respectively, π, 0, 0.5, and 2/3. The only (control) parameter that is changing is the amplitude of the driving force ϕ_0. This changes from 0.97 for the upper left to 1.2 for the lower right diagram.

The main characteristic of chaos is the *exponential* divergence of neighboring trajectories. A very nice illustration of this phenomenon for the

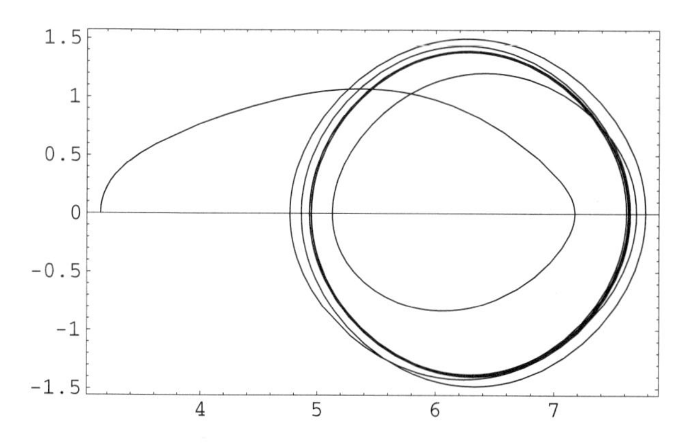

FIGURE 6.22. The moderately driven dissipative pendulum with $\gamma = 0.3$ and $\phi_0 = 0.5$. After a transient motion, the pendulum settles down into a closed trajectory.

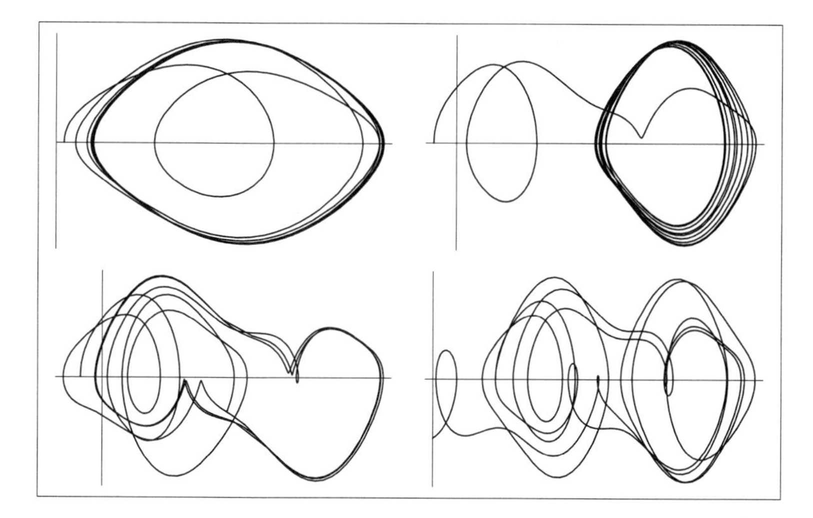

FIGURE 6.23. Four trajectories in the phase space of the damped driven pendulum. The only difference in the plots is the value of ϕ_0, which is 0.97 for the upper left, 1.1 for the upper right, 1.15 for the lower left, and 1.2 for the lower right diagrams.

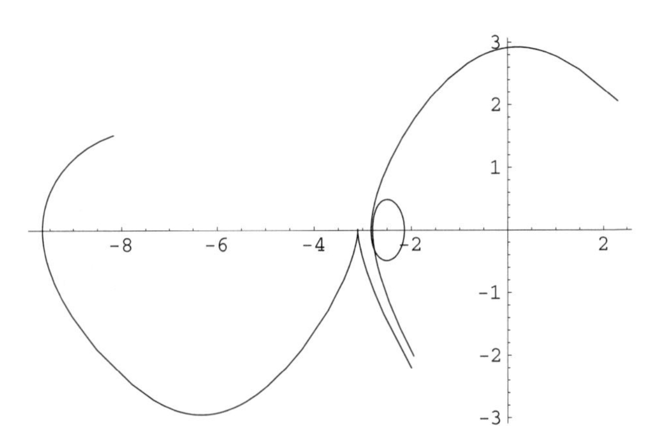

FIGURE 6.24. The projection onto the $\theta\omega$-plane of two trajectories starting at approximately the same point near $(-2, -2)$ diverge considerably after eight units of time. The loop does not contradict the DE uniqueness theorem!

nonlinear pendulum is depicted in Figure 6.24, where two nearby trajectories in the neighborhood of point $(-2, -2)$ are seen to diverge dramatically (in eight units of time).

However, something peculiar is happening here! One of the trajectories loops on itself. Why is this bad? Because of the uniqueness theorem for DEs. For our purposes this theorem states that if the dynamical variables and their first derivatives of a system are specified at some (initial) time, then the evolution of the system in time is uniquely determined. In the context of phase space this means that from any point in phase space only one trajectory can pass. If two trajectories cross, the system will have a "choice" for its further development starting at the intersection point, and the uniqueness theorem does not allow this. So why is one of the trajectories crossing itself? It is not! The plots in that figure are *projections* of the three-dimensional trajectories onto the $\theta\omega$-plane. The three-dimensional figures in which α is also included the trajectories never cross (see Figure 6.25).

6.7 Time-Independent Schrödinger Equation

Chapters 12 and 13 of *MM* discuss several special functions of mathematical physics.

Many, if not all, special functions of mathematical physics, with a wide range of applications, are solutions to eigenvalue problems, i.e., second-order DEs of the type discussed in Section 5.5. As an important paradigm of such eigenvalue problems and to gain insight into the method of solving them, we consider a few examples of the time-independent Schrödinger equation in one dimension, obtained by separating time from the (single) space coordinate in the (time-dependent) Schrödinger equation. The time-

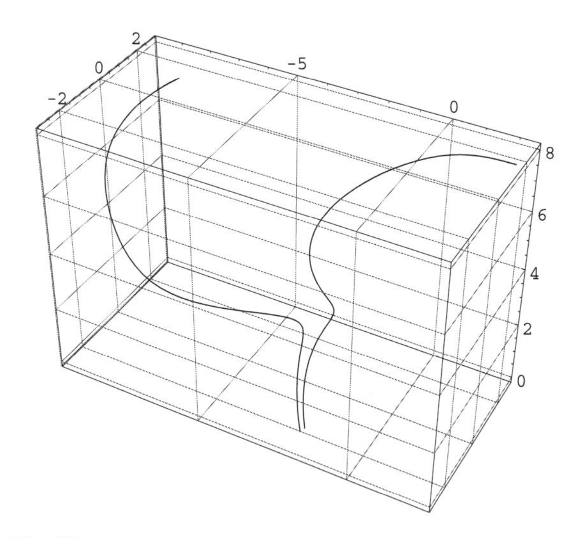

FIGURE 6.25. The two trajectories of Figure 6.24 shown in the full three-dimensional phase space.

independent Schrödinger equation is of the form

$$-\frac{\hbar^2}{2m}\frac{d^2\psi}{dx^2} + V(x)\psi = E\psi \tag{6.11}$$

where $V(x)$ and E are, respectively, the potential and total energy of the particle whose mass is m. This equation can be written as

$$\frac{d^2\psi}{dx^2} + q(x)\psi = -k^2\psi \quad \text{where} \quad q(x) = -\frac{2m}{\hbar^2}V(x), \quad k^2 = \frac{2m}{\hbar^2}E \tag{6.12}$$

which is identical to Equation (5.35).

6.7.1 *Infinite Potential Well*

As our simplest example, consider an electron in an infinite potential well described by

$$V(x) = \begin{cases} 0 & \text{if} \quad 0 < x < L \\ \infty & \text{otherwise} \end{cases}$$

Since no particle can have an infinite amount of energy, it is forbidden to go outside the region $0 < x < L$. This means that whatever the solution is for the inside, it must vanish at $x = 0$ and $x = L$. The potential energy being zero inside, the Schrödinger equation (6.12) becomes

$$\frac{d^2\psi}{dx^2} + k^2\psi = 0, \qquad k^2 = \frac{2mE}{\hbar^2} \tag{6.13}$$

Our task is to solve this simple equation by discretizing it and using Equation (5.36). But first a word on its analytic solution. The most general solution of the DE in (6.13) is

$$\psi(x) = A\cos kx + B\sin kx$$

Because $\psi(0) = 0$, A must vanish; and $\psi(L) = 0$ gives

$$B\sin kL = 0 \Rightarrow \sin kL = 0 \Rightarrow kL = l\pi \quad \text{or} \quad k = \frac{l\pi}{L}, \quad l = 1, 2, \ldots \tag{6.14}$$

where $B = 0$ has been excluded because it leads to the trivial solution $\psi(x) = 0$. This "quantization" of k leads directly to the quantization of energy by Equation (6.13). We say that the energy eigenvalues and the corresponding eigenvectors (or eigenfunctions) are

$$E_l = \frac{\hbar^2}{2m}\left(\frac{l\pi}{L}\right)^2 \qquad \psi_l(x) = B\sin\left(\frac{l\pi x}{L}\right), \quad l = 1, 2, \ldots \tag{6.15}$$

Note that B remains unspecified unless we impose some extra conditions, such as normalization, i.e., the property that the integral of the square of the eigenfunction over all values of x is one.[1] Even so, the overall sign of the eigenvectors will be arbitrary. Let us see if we can reproduce these results using Equation (5.36).

With $p(x) = 1$, $q(x) = 0$, and $\lambda = -k^2$, Equation (5.36) becomes

$$y_{i+1} - 2y_i + y_{i-1} = -h^2k^2y_i \tag{6.16}$$

Let us take L to be 1, and divide the interval $0 < x < 1$ into 10 parts with $y(0) = y_0 = 0$ and $y(1) = y_{10} = 0$. Then the nonzero y's will be y_1 through y_9, and (6.16) will consist of nine equations in nine unknowns, which can be written in matrix form as

$$\begin{pmatrix} -2 & 1 & 0 & 0 & 0 & 0 & 0 & 0 & 0 \\ 1 & -2 & 1 & 0 & 0 & 0 & 0 & 0 & 0 \\ 0 & 1 & -2 & 1 & 0 & 0 & 0 & 0 & 0 \\ 0 & 0 & 1 & -2 & 1 & 0 & 0 & 0 & 0 \\ 0 & 0 & 0 & 1 & -2 & 1 & 0 & 0 & 0 \\ 0 & 0 & 0 & 0 & 1 & -2 & 1 & 0 & 0 \\ 0 & 0 & 0 & 0 & 0 & 1 & -2 & 1 & 0 \\ 0 & 0 & 0 & 0 & 0 & 0 & 1 & -2 & 1 \\ 0 & 0 & 0 & 0 & 0 & 0 & 0 & 1 & -2 \end{pmatrix} \begin{pmatrix} y_1 \\ y_2 \\ y_3 \\ y_4 \\ y_5 \\ y_6 \\ y_7 \\ y_8 \\ y_9 \end{pmatrix} = -(0.1)^2k^2 \begin{pmatrix} y_1 \\ y_2 \\ y_3 \\ y_4 \\ y_5 \\ y_6 \\ y_7 \\ y_8 \\ y_9 \end{pmatrix} \tag{6.17}$$

MM, pp. 190–193

Equation (6.17) is a matrix eigenvalue equation, which can be solved using

[1]This arises from the fact that the probability density (here, probability per unit length) is $|\psi(x)|^2$.

Mathematica. The form of its matrix, called *tridiagonal* for obvious reasons, recurs in many applications of the Schrödinger equation.

The dimensionality of the matrix of Equation (6.17), which coincides with the number of divisions of the interval of interest, is too small for accurate numerical calculations. Ideally, we want to be able to construct matrices of variable dimensions so that we can gauge the accuracy of our calculations. The matrix of Equation (6.17) is very similar to that encountered in Section 2.6, and we can use the code in that section to construct it. However, *Mathematica* has a built-in command called `Switch` that fits in very nicely here. If we type in

using **Switch** in *Mathematica*

```
In[1]:= mInfPotential[n_]:=Table[Switch[i-j,-1,1.0,0,-2.0,
          1,1.0,_,0],{i,n},{j,n}]
```

then the command `mInfPotential[9]//MatrixForm` produces the matrix of Equation (6.17). Here `Switch` works like this: it evaluates $i-j$ in the `Table`; if the result is -1 (the second argument) it places a 1.0 (the third argument) in the ijth position; if the result is 0 (the fourth argument) it places a -2.0, and if the result is 1 (the sixth argument) it places a 1.0 there; in the rest of positions it places zeros (see below for details).

We now find the eigenvalues and eigenvectors of our tridiagonal matrix. For eigenvalues, we type in

```
In[2]:= EigVal[n_]:=(n+1)^2Eigenvalues[mInfPotential[n]]
```

The left-hand side is just the name we have given our eigenvalues with the first argument specifying which eigenvalue. The expression `Eigenvalues` calculates the eigenvalues of a matrix as a list. The factor at the beginning of the right-hand side is $1/h^2$, which cancels h^2 in (6.16), leaving us with $-k^2$.

If we type in `EigVal[100]`, *Mathematica* calculates a list of the eigenvalues of the 100×100 matrix and arranges the result in *descending* order. Let us give this list a name:

```
In[3]:= lisEigVal = EigVal[100];
```

Here we have used *immediate assignment* (=) to evaluate the right-hand side immediately, so that `lisEigVal` is now a list of 100 actual numbers available for our use. We could have used *delayed assignment* (:=), as in

immediate and delayed assignments

```
lisEigVal[n_]:= EigVal[n]
```

in which case the evaluation of the right-hand side would have been delayed until we requested it by typing in `lisEigVal[100]`. The disadvantage of the latter is that every time we request the evaluation, it recalculates the right-hand side. Sometimes that is what we want, but not when the calculation involves a list, whose members ought to be available for manipulation at different times.

As noted above, *Mathematica* arranges the eigenvalues in descending order. Normally, however, we want the lowest eigenvalue to be first. So, we Reverse the order, and pick the jth eigenvalue using Part:

use of **Reverse** in *Mathematica*

```
In[4]:= OneEigVal[j_] := Part[Reverse[lisEigVal], j]
```

For example, OneEigVal[1] gives -9.86881, which is very close to the analytic value of $-\pi^2 = -9.86960$. Similarly, OneEigVal[2] gives -39.4657, which is very close to the analytic value of $-(2\pi)^2 = -39.4784$. For low-lying states, which by definition have small eigenvalues, the numerical calculation gives very good results. However, for large eigenvalues, the agreement between numerical calculation and analytic result dwindles. For instance, OneEigVal[50] yields $-20,084.7$ as opposed to $-(50\pi)^2 = -24,674$. The reason has to do with the eigen*functions*, which are discussed below. Basically, the eigenfunctions of small eigenvalues are much smoother than those of large eigenvalues. In fact, the 50th eigenfunction crosses the x-axis 49 times between 0 and 1! For these to "appear" smooth, we have to decrease h considerably.

Let us now calculate the eigenvectors. The command

```
In[5]:= EigVec[n_]:=Eigenvectors[mInfPotential[n]]
```

finds the list of the eigenvectors, to which we assign a name

```
In[6]:= lisEigVec = EigVec[100];
```

and pick the jth one (after we Reverse the order)

```
In[7]:= OneEigVec[j_] := Part[Reverse[lisEigVec], j]
```

The components of these vectors are the y_i's. For example, to find the (components of the) first eigenvector, called the *ground state*, we type in

```
In[8]:= y[i_] := Part[OneEigVec[1], i]
```

To plot the function, we also need the x-values:

```
In[9]:= IndVar[n_] := Do[x[i]=i/(n+1), {i, 0, n+1}];
        IndVar[100]
```

This produces a list, whose first member is x[0] with value 0 and whose last member is x[101] with value 1. To plot our function, we need to pair the x's with the y's. However, the y list goes from y[1] to y[100]. But we know that y[0] and y[101] are both zero. So, we Join the zeroth and the 101st pairs to the rest, which are created by a Table command:

```
In[10]:= listOfPoints = Join[{{x[0],0}}, Table[{x[i],y[i]},
           {i,1,100}], {{x[101],0}}];
```

We now use ListPlot to plot the function,

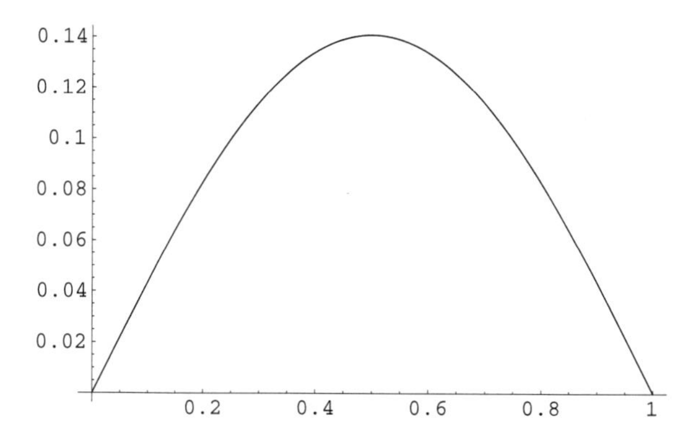

FIGURE 6.26. The lowest energy eigenfunction, or the ground state, of an infinite potential well calculated numerically.

```
In[11]:= ListPlot[listOfPoints, PlotJoined -> True,
           PlotRange -> All]
```

which yields Figure 6.26.

It is to be noted that eigenfunctions are always determined to within a multiplicative constant. Therefore, we have to compare the function of Figure 6.26 to Equation (6.15) with this arbitrariness in mind. If we choose B of (6.15) in such a way that the maxima of the two functions coincide, then they should be (very nearly) equal. The maximum of the plotted function can be obtained by typing in `y[50]`, which yields 0.140702. With $n = 1$, $B = 0.1407025$, and $L = 1$, (6.15) gives $0.1407025 \sin(\pi x)$. Let us plot the difference of the analytic and the calculated functions:

```
In[12]:= ListPlot[Table[{x[i],y[i]-0.1407025 Sin[Pi x[i]]},
           {i,1,100}], PlotJoined->True, PlotRange->All]
```

The result, shown in Figure 6.27, indicates a small difference between the analytic and the numerical solutions. This difference, of course, decreases as we increase the sample points in the interval $(0, 1)$.

The next eigenfunction can be obtained by changing `In[8]` to

```
In[13]:= y[i_] := Part[OneEigVec[2], i]
```

We then reevaluate `In[10]` and `In[11]` to obtain the first plot of Figure 6.28. We can use an input similar to `In[13]`—with 2 replaced by 3 and 4—to generate the other two plots of Figure 6.28. Readers familiar with the solutions of the infinite-well potential will recognize these as the expected eigenstates with the characteristic increase in the number of nodes (horizontal-axis crossings) with increase in n.

Before going any further, let us record the new *Mathematica* commands we have learned:

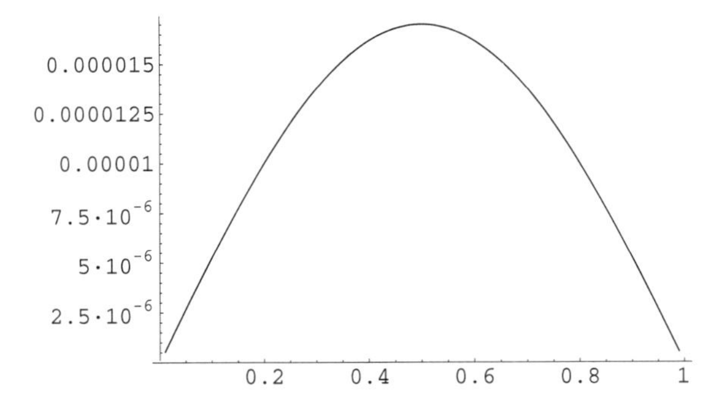

FIGURE 6.27. The difference between analytic and numerical calculation of the lowest-energy eigenfunction of an infinite potential well.

`Eigenvalues[m]`	find the list of the eigenvalues of matrix **m**
`Eigenvectors[m]`	find the list of the eigenvectors of matrix **m**
`Reverse[list]`	reverse the order of elements in the *list*
`Switch[expr,a1,v1,a2,v2, ... ,_,u]`	evaluates *expr*, then compares it with each of the a_i, in turn, evaluating and returning v_i corresponding to the first match found. After the matches are all tested, it fills the rest of ijth spots with u.

6.7.2 The General Case

Instead of going directly to the next example, let us generalize the code of the previous subsection. Thus, we translate Equation (5.36) into the *Mathematica* language by typing in

```
In[1]:= mOneDSchEq[n_,xL_,xR_] := Table[Switch[i-j,-1,
        p[x[i]],0,((xR-xL)/(n+1))^2 q[x[i]]-2.0 p[x[i]],
        1,p[x[i]],_,0], {i,n}, {j,n}]
```

where xR and xL are two (sufficiently large) values of x that define our interval of calculation. These are potential-dependent values for which ψ is so small that we can safely set it equal to zero. For the infinite potential of the previous subsection, xL was 0 (the left boundary of the well) and xR was 1 (the right boundary of the well). For other potentials they are obtained either by physical intuition, or by trial and error. The functions $p(x)$ and $q(x)$ ought to be defined in a separate statement.

Next we find the eigenvalues,

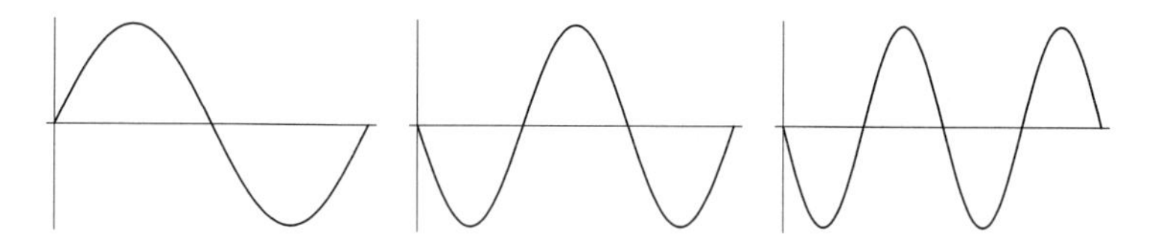

FIGURE 6.28. The eigenfunctions of the first excited state with $l = 2$ (left), the second excited state with $l = 3$ (middle), and the third excited state with $l = 4$ (right).

```
In[2]:= EigVal[n_, xL_, xR_] := ((xR-xL)/(n+1))^2
            Eigenvalues[mOneDSchEq[n, xR, xL]]
```

the defining statements for the coefficient functions, as well as assignments of values to xL and xR

```
In[3]:= q[x_] := ... ; p[x_] := ... ; xL = a; xR = b;
```

and the array of the independent variable,

```
In[4]:= Xarray[n_] := Do[x[i] = xL+i(xR-xL)/(n+1),
            {i, 0, n+1}]; Xarray[200]
```

The list of eigenvalues will then be created by a statement such as

```
In[5]:= lisEigVal = EigVal[100, a, b];
```

from which we can select the jth one:

```
In[6]:= OneEigVal[j_] := Part[Reverse[lisEigVal], j]
```

The eigenvectors are calculated similarly:

```
In[7]:= EigVec[n_, xL_, xR_] :=
            Eigenvectors[mOneDSchEq[n, xR, xL]]
```

generates the list of eigenvectors, from which

```
In[8]:= OneEigVec[j_] := Part[Reverse[lisEigVec], j]
```

picks the jth vector. These vectors are not normalized; so, as we are trying to devise a general procedure, let us include this normalization factor. The normalization condition is

$$\int_{-\infty}^{+\infty} |\psi(x)|^2 dx = 1$$

which is achieved by dividing ψ by the constant

$$N = \sqrt{\int_{-\infty}^{+\infty} |\psi(x)|^2 dx}$$

In our case, the range of integration is (xL, xR). Thus, the normalization constant becomes

$$N = \sqrt{\int_{xL}^{xR} |\psi(x)|^2 dx}$$

Using Simpson's rule of integration (4.7), we obtain

$$N = \sqrt{\frac{h}{3}\left\{4\sum_{k=1}^{n/2}(y_{2k-1})^2 + 2\sum_{k=1}^{n/2-1}(y_{2k})^2\right\}}$$

where y_0 and y_n are set equal to zero.[2] In *Mathematica* language, and for the first eigenvector, this is written as

```
In[9]:= normConst[n_] := Sqrt[(xR-xL)/(3 (n+1))
        (Sum[4.0 (Part[OneEigVec[1], 2 k-1])^2, {k,1,n/2}]
        +Sum[2.0(Part[OneEigVec[1],2 k])^2,{k,1,n/2-1}])]
```

Then, for $n = 100$,

```
In[10]:= y[i_] := Part[OneEigVec[1], i]/normConst[100]
```

produces the components of the first *normalized* eigenvector.

Finally to plot the points so found, we write the statement

```
In[11]:= listOfPoints[in_,fin_] :=
        Table[{x[i],y[i]}, {i,in,fin}];
```

where we have introduced arguments to specify the initial and final values of i in the plot. In most cases, we don't have to include all values of x, because ψ drops to very small values—indistinguishable from zero on the plots—before reaching xL and xR.

6.7.3 *Finite Potential Well*

With the general procedure at our disposal, we can look at some specific examples. The first example is a finite potential well described by

$$V(x) = \begin{cases} 0 & \text{if } 0 < x < L \\ V_0 & \text{otherwise} \end{cases}$$

We are interested only in the bound states, i.e., states whose energies are less than V_0. As in the case of the infinite potential well, only the low-lying

[2]We should really set $y_{n+1} = \psi(L)$ equal to zero. But since y_n is presumably very small, the error introduced is very small. We pick n because that is the number specified in the code, rather than $n + 1$.

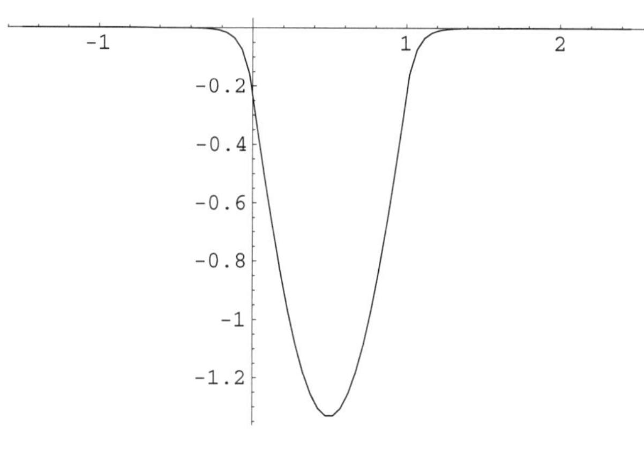

FIGURE 6.29. The normalized eigenfunction of the ground state of an electron in a finite potential well with height 10 eV.

states are expected to be sufficiently accurate, and we shall concentrate on those. Of course, by making h small enough (or the number of sample points large enough), we obtain better accuracy for the higher eigenvalues.

We shall be using atomic units, and take L to be 1 nm (10^{-9} m). The height of the potential is measured in electron volts, or eV, which is equal to 1.6×10^{-19} Joule. Let us take V_0 to be 10 eV. Then, for an electron, with a mass of 9.1×10^{-31} kg, we have

$$q(x) = \begin{cases} 0 & \text{if} \quad 0 < x < 1 \\ -2mV_0/\hbar^2 = -255.5 \text{ eV}^{-1} \cdot \text{nm}^{-2} & \text{otherwise} \end{cases}$$

It follows that the input line `In[3]` of the previous subsection should be written as

```
In[3]:= q[x_] := If[x>0 && x<1, 0, -255.5]; p[x_]:=1.0;
        xL=-2; xR=3;
```

The succession of the other input lines then produces -7.78481 for $-k^2$ and Figure 6.29 for the normalized ground-state eigenfunction. The reason the the plot has fallen below the x-axis is the arbitrariness in the overall sign of the function. *Mathematica* has picked a negative sign for the ground state.

It is worthwhile to point out that we have obtained the solution without imposing any *extra* boundary conditions. In analytic solutions, not only do we demand the vanishing of ψ at $\pm\infty$ (equivalent to its vanishing at `xL` and `xR` in numerical calculation), but also the continuity of both ψ and its derivative at the boundaries of the well, $x = 0$ and $x = L$. This is because of the discontinuity of the potential at the boundary. In discrete calculations, the notion of continuity does not exist, because one can always construct

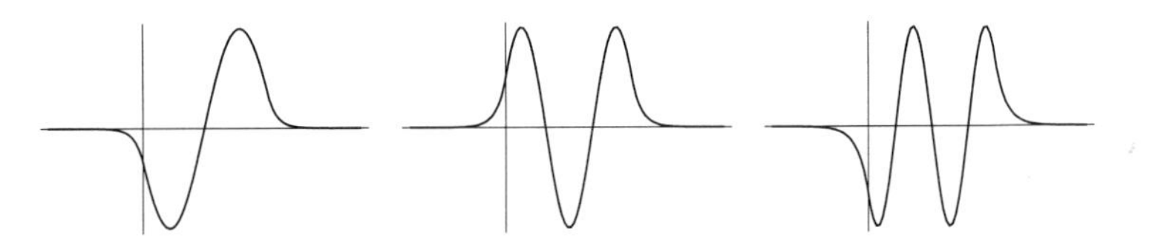

FIGURE 6.30. The normalized eigenfunctions of the first (left), second (middle), and third (right) excited states of an electron in a finite potential well with height 10 eV.

continuous potentials that take the value (almost) zero for all the discrete points inside the well and the value (almost) V_0 for all the discrete points outside.

How good is the numerical calculation? To find out, let us analytically calculate the ground-state eigenvalue. First, we divide the x-axis into three regions, left, inside, and right, with respective wave functions, ψ_l, ψ_{in}, and ψ_r. With

$$\kappa = \sqrt{\frac{2m}{\hbar^2}(V_0 - E)} \quad \text{and} \quad k = \sqrt{\frac{2mE}{\hbar^2}}$$

and the infinity boundary conditions, the three ψ's will reduce to

$$\psi_l(x) = Ae^{\kappa x}, \quad \psi_{in}(x) = B\cos kx + C\sin kx, \quad \psi_r(x) = De^{-\kappa x}$$

Imposing the continuity of ψ and its derivative at the two boundaries of the well (at $x = 0$ and $x = L$) gives the equation

$$\kappa = \frac{k\sin kL - \kappa\cos kL}{\cos kL + (\kappa/k)\sin kL}$$

Substituting the values for an electron in a 1-nm-wide potential of height 10 eV, the preceding equation gives (after a little algebra)

$$\tan k = \frac{-2k\sqrt{255.5 - k^2}}{255.5 - 2k^2}$$

This is a transcendental equation that can be solved only numerically. Typing in

```
FindRoot[Tan[k]+2 k Sqrt[255.5-k^2]/(255.5-2 k^2)==0,
          {k, 2.6}]
```

yields $\{x \to 2.79062\}$, which compares very favorably with our numerical calculation of $\sqrt{7.78481} = 2.79013$.

The first few excited states of the finite potential can also be solved by changing a few parameters. Figure 6.30 shows the first three excited states. For better accuracy, we took the number of divisions to be 200.

One can go to the limit of an infinite potential well by taking V_0 to be very large. How large? It turns out that to obtain the same ground-state eigenvalue—to two decimal places—as the infinite potential, one has to choose a V_0 of 10^8 eV!

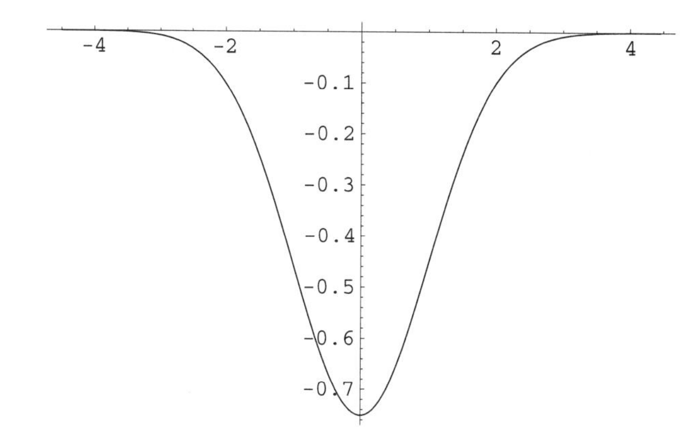

FIGURE 6.31. The normalized eigenfunction of the ground state of a harmonic oscillator.

6.7.4 Harmonic Oscillator

One of the most recurrent potentials in quantum theory is the harmonic oscillator potential. This is because, to the first approximation, all potentials look like a harmonic oscillator at their equilibrium point. With $V(x) = \frac{1}{2}kx^2$ and $\omega = \sqrt{k/m}$, Equation (6.12) will have $q(x) = -(m\omega/\hbar)^2 x^2$. For simplification, we take the numerical value of ω to be $\hbar/m$. Then, Equation (6.12) becomes

$$\psi''(x) + q(x)\psi(x) = -k^2\psi(x) \quad \text{where} \quad q(x) = -x^2 \quad \text{and} \quad k^2 = \frac{2E}{\hbar\omega} \tag{6.18}$$

The numerical solution of a quantum harmonic oscillator is obtained by appropriate changes to the inputs of Section 6.7.2. In particular, for `In[3]` we have

```
In[3]:= q[x_] := -x^2; p[x_]:=1.0; xL=-5; xR=5;
```

where we have chosen -5 and $+5$ for the endpoints of our interval. We then execute the input lines of the eigenvalue evaluation, and get -0.999845, -2.99923, and -4.99799 for the first three eigenvalues ($-k^2$). Equation (6.18) then implies the energies of the first three lowest-lying states are $0.4999225\hbar\omega$, $1.499615\hbar\omega$, and $2.498995\hbar\omega$, respectively. These are to be compared with the respective exact results of $\frac{1}{2}\hbar\omega$, $\frac{3}{2}\hbar\omega$, and $\frac{5}{2}\hbar\omega$. It is clear that, even with a meager 200 sample points, the procedure gives a reasonably accurate result.

What about the eigenfunctions? By inserting 1, 2, and 3 for the argument of `OneEigVec[ ]` in appropriate places, we obtain the ground state (Figure 6.31) and the first two excited states (Figure 6.32) of the harmonic oscillator. The reader—bearing in mind the overall arbitrariness in sign—

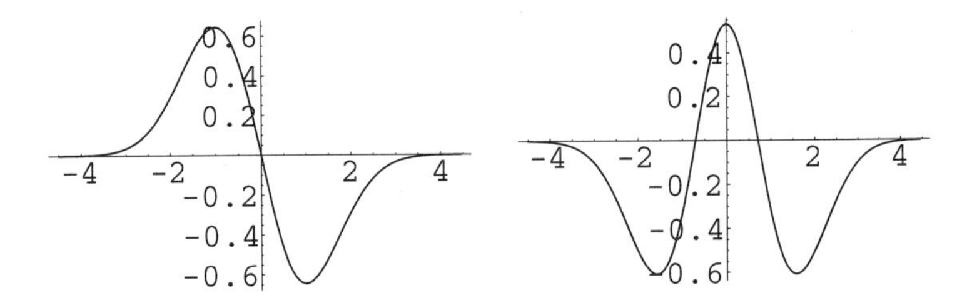

FIGURE 6.32. The normalized eigenfunctions of the first (left) and second (right) excited states of a harmonic oscillator.

may wish to compare these plots with those of the analytic solutions,

$$\psi_0(x) = \frac{e^{-x^2/2}}{\pi^{1/4}}, \quad \psi_1(x) = \frac{2xe^{-x^2/2}}{\sqrt{2}\,\pi^{1/4}}, \quad \text{and} \quad \psi_2(x) = \frac{(2x^2-1)e^{-x^2/2}}{\sqrt{2}\,\pi^{1/4}}$$

to determine the accuracy of our procedure for eigenfunction calculation.

In some applications, anharmonic oscillators become important. This is because in a Taylor expansion of a potential, one may want to keep higher orders beyond the harmonic oscillator term for a more accurate representation of the potential. One term that is sometimes considered is a quartic term. So, let us add to our harmonic oscillator potential a (smaller) term of the fourth power in x and opposite in sign. Then we have $q(x) = -x^2 + 0.1x^4$. Choosing $xL = -8$, $xR = 8$, and $n = 300$, and executing the appropriate codes of Section 6.7.2, we obtain a ground-state eigenvalue of 0.896746 and the plot of Figure 6.33 for the eigenfunction of the ground state. At first, the graph looks strange, but upon a little reflection, it emerges as feasible. Here is why. When x is small, the dominant term of the potential is $-x^2$, and we expect the function to look like the ground state of the simple harmonic oscillator; and it does. Between $x = \pm 2$ and $x = \pm 3.5$, the graph shows a (exponential) decay, indicating some kind of potential barrier. Beyond $x = \pm 4$, we see an oscillatory behavior indicative of a free particle.

The graph of the potential in Figure 6.34 makes the claim above very clear. Close to the origin, the figure is very similar to a parabola. The two humps on either side act as (finite) barriers, through which the particle can "tunnel," decaying (and losing amplitude) in the process. Once it gets to just over $x = 3$, the particle's energy is larger than the potential energy, and it is no longer bound. Like all free particles, it now oscillates with a wavelength determined by the difference between E and the potential energy. As it moves farther away from the origin, this difference increases, resulting in shorter wavelengths.

Other eigenvalues and eigenfunctions of the anharmonic oscillator can also be calculated. For example, the next eigenvalue will turn out to be -2.06656, whose corresponding eigenfunction is shown in Figure 6.35.

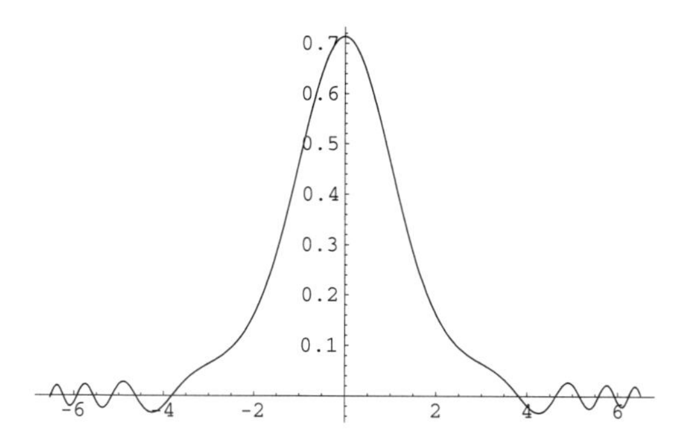

FIGURE 6.33. The normalized eigenfunction of the ground state of the anharmonic oscillator of Figure 6.34.

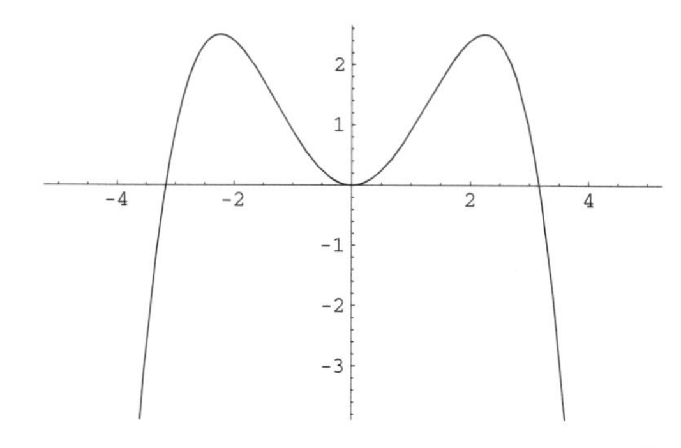

FIGURE 6.34. The anharmonic oscillator potential $x^2 - 0.1x^4$.

6.8 Problems

Problem 6.1. Experiment with a one-dimensional projectile by varying the drag coefficient a and its decay rate b [see Equation (6.2)]. Then plot and interpret the result. Make sure you vary these two parameters over a wide range to see some noticeable changes in the graphs.

Problem 6.2. The moon has a mass of 7.35×10^{22} kg and a radius of 1.74×10^6 m. Consider a one-dimensional projectile on the moon fired with initial velocity v_0.
(a) By assigning values to v_0 and plotting the resulting plot [obtained by solving Equation (6.2) with $a = 0$], estimate the escape velocity (the smallest speed for which the projectile will not return) of the moon.
(b) If the moon had an atmosphere with $a = 0.2$ and $b = 0.0002$, what would its escape velocity be?

Problem 6.3. Change Equation (6.5) in such a way that the drag force (at height zero) is proportional to the square of velocity, so that the *magnitude* of the force is av^2 rather than av. Now write a *Mathematica* code to solve

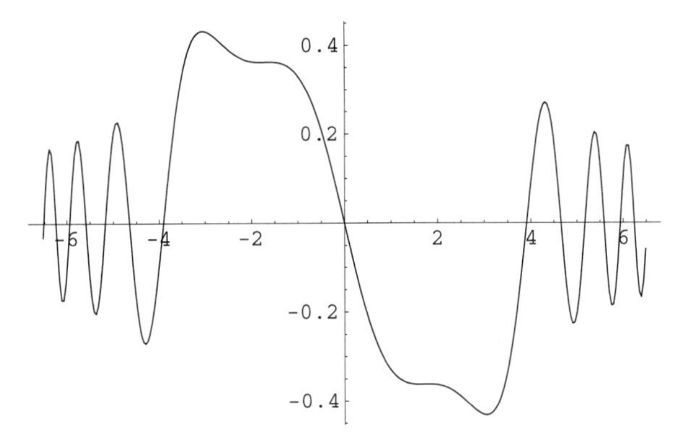

FIGURE 6.35. The normalized eigenfunction of the first excited state of the anharmonic oscillator of Figure 6.34.

the new set of DEs. Plot the solution for your choice of the values of the parameters. Hint: The *vector* multiplying the exponential can be written as $av\mathbf{v}$.

Problem 6.4. By evaluating the two values of y at $T/2$, find the semimajor axes of the bound orbits of Figure 6.10 and see if Kepler's third law, $T^2 = ka^3$, holds.

Problem 6.5. By comparing the location of the moon at $t = 0$ and at $t = T/2$, where T is the period of the moon, find the distance between the two foci, and show that it is not zero. See Figure 6.12 and the discussion of the two-body problem surrounding it.

Problem 6.6. Find the path of a particle of unit charge in a magnetic field given by $B = 2z$ pointing in the positive x-direction (see Problem 5.5 for the relevant DEs). Pick initial conditions of your choice for the particle.

Problem 6.7. The **Duffing oscillator** is an oscillator obeying the nonlinear DE

$$\ddot{x} + 2\gamma\dot{x} + x + x^3 = a\cos(\omega_D t).$$

(a) Employing the trick used in Equation (6.10), turn the Duffing equation into a set of first-order DEs.
(b) For $\gamma = 0.1$, $a = 3$, and $\omega_D = 4$, find the solution of this equation, and plot it for $0 \leq t \leq 30$.
(c) Using part (b) and `ParametricPlot`, make a phase-space diagram of this oscillator.
(d) With $\gamma = 0.1$ and $\omega_D = 0.1$, solve the Duffing equation for different values of a, especially large values, and plot the phase-space trajectory of the oscillator.

Problem 6.8. In the Lorenz equation of Problem 5.4, let $\sigma = 10$, $r = 3$, and $q = 8/3$.

(a) Solve the equation, and make a parametric plot of z versus y for $0 \leq t \leq 9$ with the initial condition $x(0) = 0$, $y(0) = 1$, and $z(0) = 0$.
(b) Now change the initial conditions to $x(0) = 0$, $y(0) = 1$, $z(0) = 1$, and make a three-dimensional parametric plot of the system.

Problem 6.9. Consider the electron in a finite potential well of depth V_0. Obtain solutions of the Schrödinger equation for larger and larger values of V_0. For what value of V_0 is the eigenvalue equal to the infinite potential case to within two decimal places?

Problem 6.10. Compare the analytic eigenfunctions for the lowest three states of a harmonic oscillator of Section 6.7.4 with those obtained numerically.

Problem 6.11. A one-dimensional potential $V(x)$ is infinite for $x \leq 0$ and is equal to $2x$ for $x > 0$. Using the technique of Section 6.7.2, find the lowest two eigenvalues and the plot of their corresponding eigenfunctions.

References

[Call 85] Callen, H. *Thermodynamics and an Introduction to Thermostatics*, New York: Wiley, 1985.

[Cran 96] Crandall, R. *Topics in Advanced Scientific Computation*, New York: Springer-Verlag/TELOS, 1996.

[Gass 98] Gass, R. *Mathematica for Scientists and Engineers*, Englewood Cliffs, NJ: Prentice Hall, 1998.

[Hass 00] Hassani, S. *Mathematical Methods: For Students of Physics and Related Fields*, New York: Springer-Verlag, 2000.

[Hass 99] Hassani, S. *Mathematical Physics, A Modern Introduction to Its Foundations*, New York: Springer-Verlag, 1999.

[Wolf 96] Wolfram, S. *Mathematica*, Cambridge, U.K.: Cambridge University Press, 1996.

Index